ELEMENTS OF ELECTRICITY

MAXWELL PRESS

ELEMENTS OF ELECTRICITY

BY

Robert Andrews Millikan, Ph.D., Sc.D.

Winner of Nobel Prize for Physics

Professor of Physics, University of Chicago Co-Author of "First Course in PHysics" Author of "Mechanics, Molecular Physics, and Heat" And Co-Author of "Electricity, sound, and Light" Member, American Physical Society

And

E.S.Bishop, Ph.D.

Instructor in Physics, University High School
University of Chicago

MAXWELL PRESS

Chennai Tirunelveli NewDelhi

MAXWELL PRESS

ISBN 978-93-90063-40-6

Printed and bound in India
MJP 886

Maxwell Press
No. 44, Nallathambi Street,
Triplicane, Chennai 600 005

Publisher : C. Janarthanan

Publisher's Note

The legacy of a country is in its varied cultural heritage, historical literature, developments in the field of economy and science. The top nations in the world are competing in the field of science, economy and literature. This vast legacy has to be conserved and documented so that it can be bestowed to the future generation. The knowledge of this legacy is slowly getting perished in the present generation due to lack of documentation.

Keeping this in mind, the concern with retrospective acquiring of rare books has been accented recently by the burgeoning reprint industry. Maxwell Press is gratified to retrieve the rare collections with a view to bring back those books that were landmarks in their time.

In this effort, a series of rare books would be republished under the banner, "Maxwell Press". The books in the reprint series have been carefully selected for their contemporary usefulness as well as their historical importance within the intellectual. We reconstruct the book with slight enhancements made for better presentation, without affecting the contents of the original edition.

Most of the works selected for republishing covers a huge range of subjects, from history to anthropology. We believe this reprint edition will be a service to the numerous researchers and practitioners active in this fascinating field. We allow readers to experience the wonder of peering into a scholarly work of the highest order and seminal significance.

MAXWELL PRESS

INTRODUCTION

THE subject of electricity is so fascinating and it covers so many important and interesting applications that a study of the laws under which this mysterious force moves is not only attractive but is fundamental in its character. Since the beginnings of our electrical knowledge the great minds of science have struggled to determine the nature and origin of electricity but the problem is not yet solved. We know, however, how electricity behaves and we can harness it and make it do our will as is evidenced by the fact that our electric lights burn with perfect reliability; our street and interurban cars carry thousands of people day after day; our countless motors turn the wheels of our factories week in and week out, while our telephones and telegraphs perform their wonderful service in our business and social lives without interruption.

¶ This work is not intended as an unsystematic and popularized treatise on electricity but is worked out in a thorough and careful manner. Many problems are given under the various topics and copious illustrations have been used to make the points clear. In order to show how the various laws and principles have been applied to our everyday life, practical applications have been made throughout the treatise such as the commercial current measuring instruments, telegraphy both wire and wireless, X-Ray and radioactivity. Another unique feature of the book is the elementary presentation of the laws of the alternating-current circuit.

¶ The authors are authorities in the electrical field, not only from the scientific but also from the teaching standpoint. It is the hope of the publishers that the volume will prove valuable as a text and as a source of interesting information for the general reader.

CONTENTS

BEGINNINGS OF ELECTRICITY

CONTENTS

MEASUREMENT AND APPLICATIONS OF THE ELECTRIC CURRENT

CONTENTS

INDUCED CURRENTS AND ELECTRIC POWER

ELEMENTS OF ELECTRICITY AND MAGNETISM*

PART I

BEGINNINGS OF ELECTRICITY

MAGNETISM

1. Natural and Artificial Magnets. It has been known for many centuries that some specimens of the ore known as magnetite (Fe_3O_4) have the property of attracting small bits of iron and steel, Fig. 1. This ore probably received its name from the fact that it is abundant in the province of Magnesia in Thessaly, although the Latin writer *Pliny* says that the word magnet is derived from the name of the Greek shepherd Magnes, who, on the top of Mount Ida, observed the attraction of a large stone for his iron crook. Pieces of ore which exhibit this attractive property for iron or steel are known as *natural magnets.*

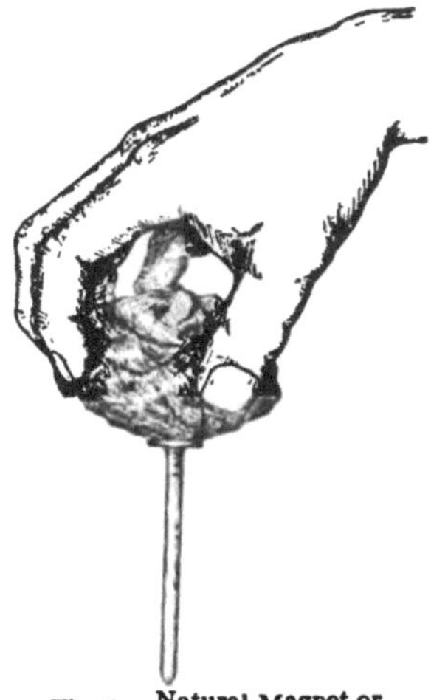

Fig. 1. Natural Magnet or Lodestone

It was also known to the ancients that artificial magnets may be made by stroking pieces of steel with natural magnets, but it was not until the twelfth century that the discovery was made that a suspended magnet would assume a north-and-south position. Because of this property, natural magnets came to be known as lodestones (leading stones); and magnets, either artificial or natural, began to be used for determining directions. The first mention

---

*This article is a modification and extension of the treatment of *Magnetism and Electricity* found in Millikan and Gale's "First Course in Physics" (Ginn & Co., Boston), to which the student is referred for additional problems and applications.

of the use of a compass in Europe is in 1190. It is thought to have been introduced from China.

Artificial magnets are now made either by repeatedly stroking a bar of steel, first from the middle to one extremity with one of the ends, or *poles*, of a magnet, and then from the middle to the other extremity with the other pole; or else by passing electric currents about the bar in a manner to be described later. The form shown in Fig. 2 is called a *bar* magnet, that shown in Fig. 3 is a *horseshoe* magnet.

Fig. 2. Bar Magnet

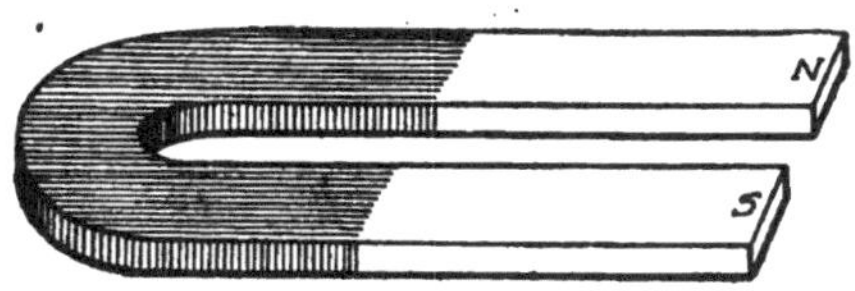

Fig. 3. Horseshoe Magnet

2. Poles of a Magnet. If a magnet is dipped into iron filings, the filings are observed to cling in tufts near the ends, but scarcely at all near the middle, Fig. 4. These places near the ends of the magnet, in which its strength seems to be concentrated, are called the *poles* of the magnet. It has been decided to call the end of a freely suspended magnet which points to the north, the *north-seeking* or *north pole*, and it is commonly designated by the letter N. The other end is called the *south-seeking* or *south pole*, and is designated by the letter S. The direction in which the compass needle points is called the *magnetic meridian*.

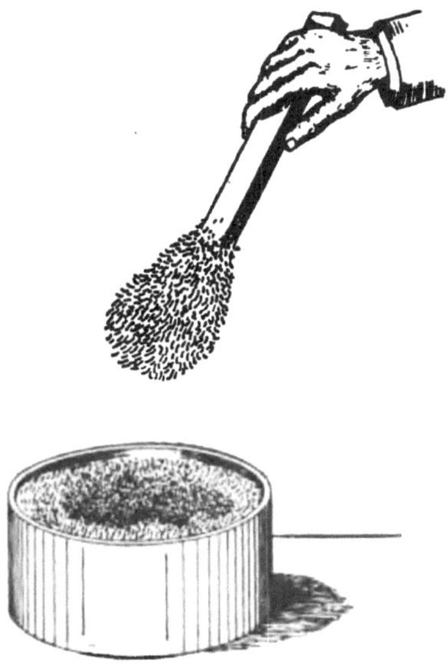

Fig. 4. Location of Poles of a Magnet

3. Laws of Magnetic Attraction and Repulsion. In the experiment with the iron filings, no particular difference was observed between the action of the two poles. That there is a difference, however, may be shown by experimenting with two magnets, either of which may be suspended, Fig. 5. If two N poles are brought near each other, each is found to repel the other. The S poles likewise are found to act in the same way. But the N pole of one magnet is found to be attracted by the *S* pole of the other. The results of these experiments may be summarized in a general law: *Magnet poles of like kind repel each other, while poles of unlike kind attract.*

This force of attraction or repulsion between poles is found, like gravitation, to vary inversely as the square of the distance between the poles; that is, separating two poles to twice their original distance reduces the force acting between them to one-fourth its original value, separating them to three times their original distance, reduces the force to one-ninth its original value, etc.

4. Magnetic Pole of Unit Strength. A pole is said to be *of unit strength* when it will repel with a force of *one dyne*, an exactly *equal* and *similar* pole placed one *centimeter* away in air. The number of units of magnetism in any pole is, then, equal to the number of dynes of force which it exerts upon a unit pole placed 1 cm. from it. Thus, if this force is found to be 100 dynes we say that the pole contains 100 units of magnetism, etc.

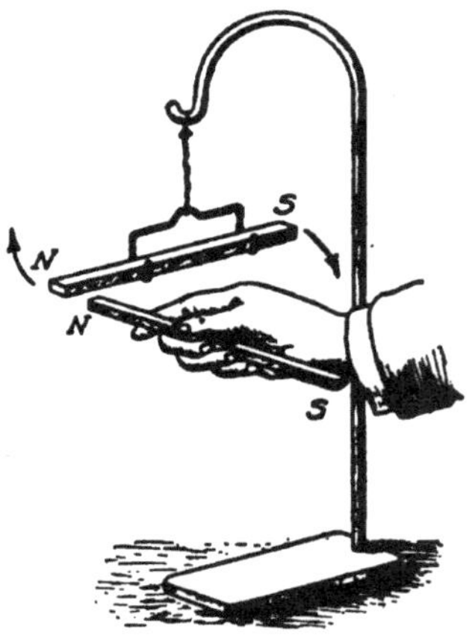

Fig. 5. Experiment Proving the Law of Magnetic Attraction and Repulsion

If, now, we were to place a pole containing 2 units of magnetism, 1 cm. from the pole containing 100 units of magnetism, the force acting between them would be found to be 200 dynes; i.e., the force of attraction or repulsion between magnetic poles is directly proportional to the product of the strengths of the two poles.

This force *F*, directly proportional to the product of the pole strengths P_1 and P_2 of the two poles and inversely proportional to the square of the distance *d* between the poles, may then be expressed by the equation

$$F = \frac{P_1 P_2}{d^2}$$

Example. An *N* pole of strength 150 is placed 25 centimeters from an *S* pole of strength 200. With what force do they act upon each other?

$$F=\frac{150\times200}{25^2}=48 \text{ dynes (force of attraction)}$$

Ans. 48 dynes

5. Magnetic Substances. Iron and steel are the only common substances which exhibit magnetic properties to a marked degree. Nickel and cobalt, however, are also attracted appreciably by strong magnets. Bismuth, antimony, and a number of other substances are actually repelled instead of attracted, but the repulsion is very small. Until quite recently, iron and steel were the only substances whose magnetic properties were sufficiently strong to make them of any value as magnets. Recently, however, it has been discovered that it is possible to make rather strongly magnetic alloys out of non-magnetic materials. For example, a mixture of 65 per cent copper, 27 per cent manganese, and 8 per cent aluminum is rather strongly magnetic. These metals are known as the *Heussler alloys*.

6. Magnetic Induction. If a small unmagnetized nail is suspended from one end of a bar magnet, it is found that a second nail may be suspended from the first nail, which itself acts like a magnet, a third from the second, etc., as shown in Fig. 6. But if the bar magnet is carefully pulled away from the first nail, the others will instantly fall away from one another, thus showing that the nails are strong magnets only so long as they are in contact with the bar magnet. Any piece of soft iron may be thus magnetized *temporarily* by holding it in contact with a permanent magnet. Indeed, it is not necessary that there be actual contact, for if a nail is simply brought near to the permanent magnet, the former is found to become a magnet. This may be proved by presenting some iron filings to one end of a nail held near a magnet in the manner shown in Fig. 7. Even inserting a plate of glass, or of copper, or of any other material except iron between S and N will not change appreciably the number of filings which cling to the end of S'. But as soon as the permanent magnet is removed, most of the filings will fall. *Magnetism produced in this way by*

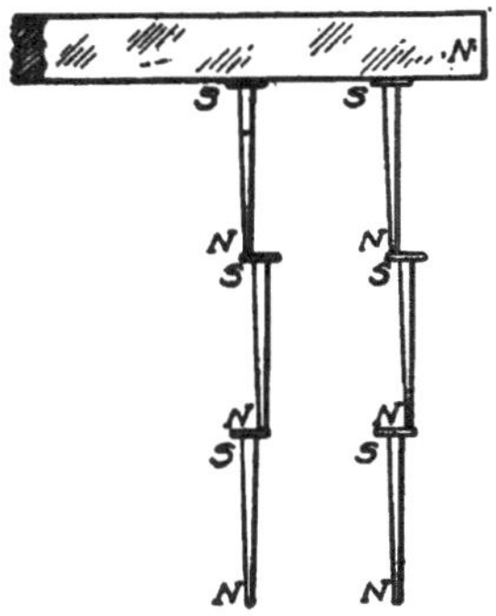

Fig. 6. Experiment Illustrating Transmission of Magnetism

the mere presence of adjacent magnets, with or without contact, is called induced magnetism. If the induced magnetism of the nail in Fig. 7 is tested with a compass needle, it is found that the *remote* induced pole *S'* is of the same kind as the inducing pole *S*, while the *near* pole *N* is of unlike kind. This is the general law of magnetic induction

Magnetic induction explains the fact that a magnet attracts an unmagnetized piece of iron, for it first magnetizes the iron by induction, so that the near pole is unlike the inducing pole, and the remote pole like the inducing pole, and then, since the two unlike poles are closer together than the like poles, the attraction overbalances the repulsion and the iron is drawn toward the magnet. Magnetic induction also explains the formation of the tufts of iron filings shown in Fig. 4, each little filing becoming a temporary magnet such that the end which points toward the inducing pole is unlike this pole, and the end which points away from it is like this pole. The bush-like appearance is due to the repulsive action which the outside free poles exert upon each other.

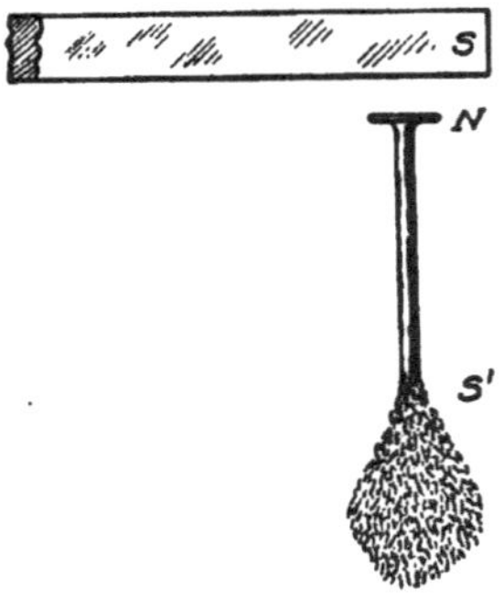

Fig. 7. Experiment Illustrating Magnetic Induction

7. Retentivity and Permeability. A piece of soft iron will very easily become a strong temporary magnet, but when removed from the influence of the magnet, it loses practically all of its magnetism. On the other hand, a piece of steel will not be so strongly magnetized as the soft iron, but it will retain a much larger fraction of its magnetism after it is removed from the influence of the permanent magnet. This power of resisting either magnetization or demagnetization is called *retentivity.* Thus, steel has a much greater retentivity than wrought iron, and, in general, the harder the steel the greater its retentivity.

A substance which has the property of becoming strongly magnetic under the influence of a permanent magnet, whether it has a high retentivity or not, is said to possess *permeability* in large degree. Thus iron is much more permeable than nickel. Permeability is measured by the amount of magnetization which a substance is able to receive; while retentivity is measured by the tenacity with which it holds this magnetization.

8. Magnetic Lines of Force. If we could separate the *N* and *S* poles of a small magnet so as to obtain an independent *N* pole, and were to place this *N* pole near the *N* pole of a bar magnet, it would move over to the *S* pole of the bar magnet, along a curved path similar to that shown in Fig. 8. The reason that the motion is along a curved rather than along a straight path is that the free pole is at one and the same time repelled by the *N* pole of the bar magnet and attracted by its *S* pole, and the relative strengths of these two forces are continually changing as the relative distances of the moving pole from these two poles are changed.

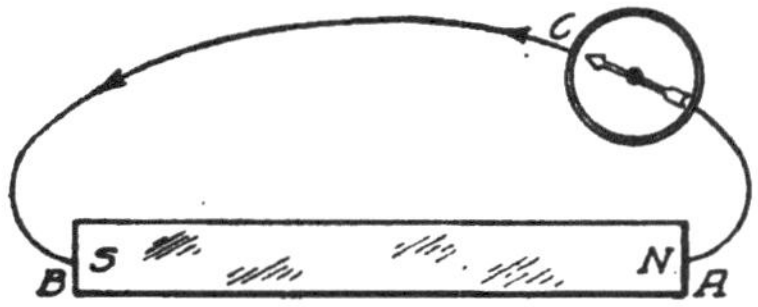

Fig. 8. Plotting Lines of Force in a Magnetic Field

It is not difficult to test this conclusion experimentally. Thus, if a bar or a horseshoe magnet is placed just beneath a flat dish containing water, Fig. 9, and a cork carrying a magnetized needle is placed near the *N* pole in the manner shown in the figure, the cork will actually be found to move in a curved path from the *N* pole around to the *S* pole of the bar magnet. In this case the cork and the needle actually move as would an independent pole, since the upper pole of the needle is so much farther from the magnet than the lower pole that the influence of the former on the motion is very small.

Any path which an independent *N* pole would take in going from *N* to *S* is called a *line of magnetic force.* The simplest way of finding the direction of this path at any point near a magnet is to hold a compass needle at the point considered, for the needle must obviously set itself along the line in which its poles would move if independent, that is, along the line of force which passes through the given point, *C*, Fig. 8.

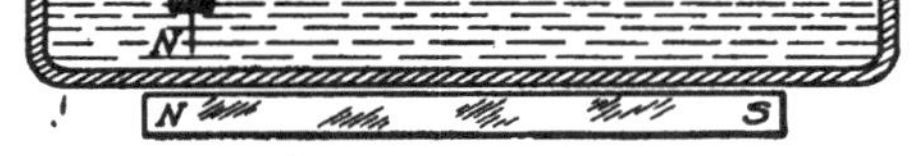

Fig. 9. Experimental Proof of Magnetic Action along Lines of Force

9. Magnetic Fields of Force. The region about a magnet in which its magnetic forces can be detected is called its *field of force.* The simplest method to gain an idea of the way in which the lines of force are arranged in the magnetic field about any magnet is to sift iron filings upon a piece of paper placed immediately over the magnet. Each little filing becomes a temporary magnet by induction and therefore, like the compass needle, sets itself in the direction of the line of force at that point. Fig. 10 shows the shape of the *magnetic field* about a bar magnet. Fig. 11 shows the direction of the *lines of force* about a horseshoe magnet. Fig. 12 is the ideal diagram corresponding to Fig. 10 and showing the lines of force emerging from the *N* pole and passing around in curved lines to the *S* pole. This way of imagining the lines of force to be closed curves passing on the outside of the magnet from *N* around to *S*, and on the inside of the magnet from *S* back to *N*, was introduced by Faraday about

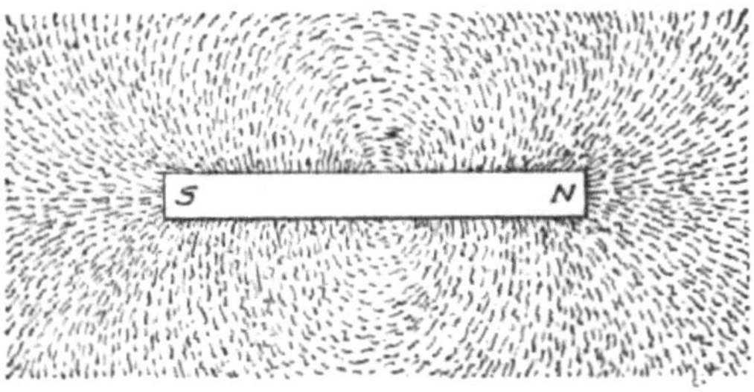

Fig. 10. Magnetic Field of Bar Magnet

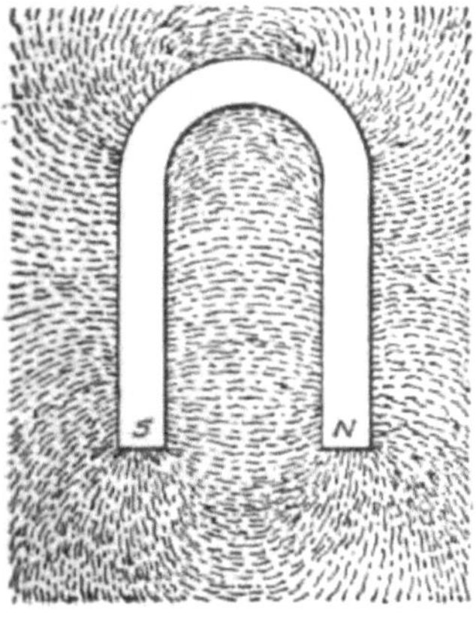

Fig. 11. Magnetic Field of Horseshoe Magnet

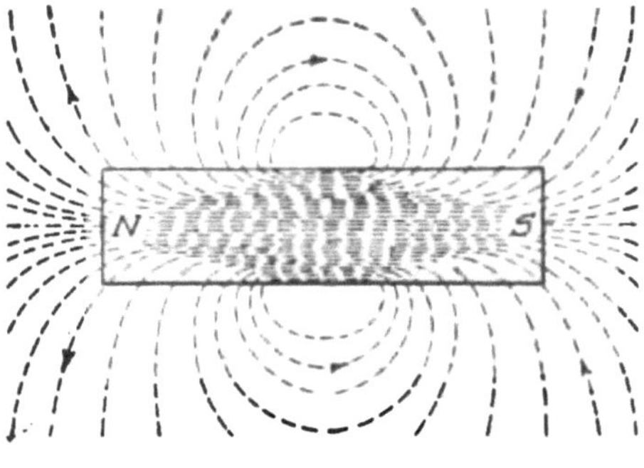

Fig. 12. Ideal Magnetic Field for Bar Magnet

1830, and has been found of great assistance in correlating the facts of magnetism.

10. Strength of a Magnetic Field. The strength of a magnetic field at any point near a magnet is defined as the number of dynes of force which a unit magnet pole would experience at the point considered. Thus, if at some point between N and S of a horseshoe magnet, Fig. 13, a unit N pole were pushed by the field from N toward S with a force of one dyne, then the magnetic field at that point would be a "field of unit strength", or a "unit magnetic field". In electrical engineering, such a field strength would be called one *gauss*. If the unit N pole were pushed from N toward S with a force of 1,000 dynes, then the field strength would be one of 1,000 gausses, etc.

If we wish to represent graphically a field of unit strength, i.e., 1 gauss, we draw one line per square centimeter through a surface such as $ABCD$, taken at right angles to the lines of force. A field of strength 2, i.e., 2 gausses, would be represented by two lines per square centimeter, a field strength of n gausses, by n lines per square centimeter, etc.; i.e., field strengths are represented by the number of lines of force drawn to the square centimeter.

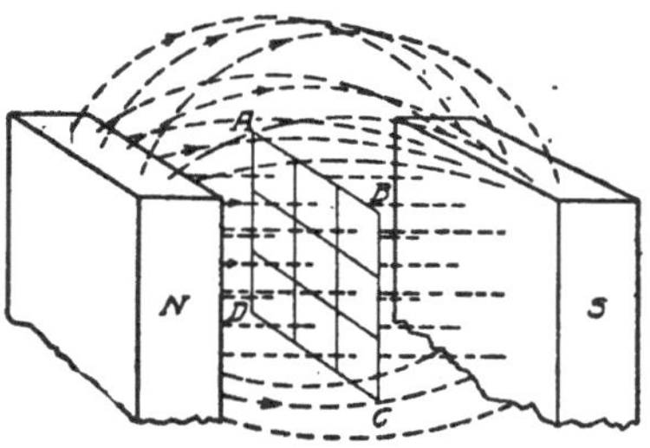

Fig. 13. Diagram Illustrating Method of Measuring Strength of Magnetic Field

11. Molecular Nature of Magnetism. If a small test tube full of iron filings be stroked from one end to the other with a magnet, it will be found to behave toward a compass needle as if it were itself a magnet, but it will lose its magnetism as soon as the filings are shaken up. If a magnetized needle is heated red hot, it is found to lose its magnetism completely. Again, if any magnet is jarred, hammered, or twisted, the strength of its poles, as measured by their ability to pick up tacks or iron filings, is found to be greatly diminished.

These facts point to the conclusion that magnetism has something to do with the arrangement of the molecules, since causes which violently disturb the molecules of the magnet weaken its magnetism. Again, if a magnetized needle is broken, each part will be found to be a complete magnet. That is, two new poles will appear at the point of breaking, a new N pole on the part which has the original S pole, and a new S pole on the part which has the original N pole. The subdivision may be continued indefinitely, but always with the same result, as indicated in Fig. 14.

This points to the conclusion that the molecules of a magnetized bar are themselves little magnets arranged in rows having their opposite poles adjacent.

If an unmagnetized piece of hard steel is pounded vigorously while it lies between the poles of a magnet, or if it is heated to redness and then allowed to cool in this position, it will be found to have become magnetized. The pounding, or heating, which is nothing but molecular bombardment, aids all of the little molecular magnets to turn and point in the direction of the magnetic field, just as tapping a compass needle supported on a dull pivot aids the compass in taking up the direction of the field. This points to the conclusion that the molecules of the steel are magnets even when the bar as a whole is not magnetized, and that magnetization consists in causing these molecular magnets to arrange themselves in rows, end to end.

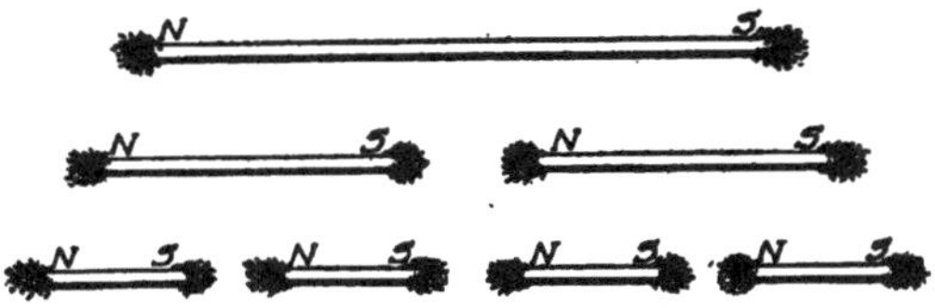

Fig. 14. Magnetic Nature of Fragments of Bar Magnet

Fig. 15. Molecular Nature of Magnetism. Condition of Molecules in Unmagnetized Bar

In an unmagnetized bar of iron or steel, then, it is probable that the molecules themselves are tiny magnets which are arranged either haphazard or in little closed groups or chains, as in Fig. 15, so that, on the whole, opposite poles neutralize each other throughout the bar. But when the bar is brought near a magnet, the molecules are swung around by the outside magnetic force into some such arrangement as that shown by Fig. 16, in which the opposite poles completely neutralize each other only in the middle of the bar. According to this view, the reason that heating and jarring weaken a magnet is that disturbances of this sort tend to shake the molecules out of alignment. On the other hand,

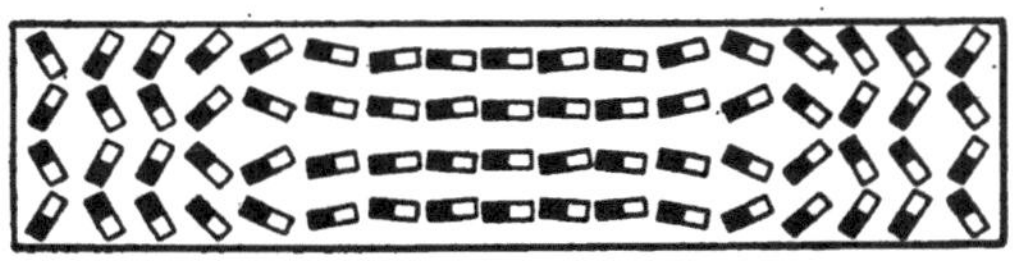
Fig. 16. Condition of Molecules of Bar after Being Magnetized

heating and jarring facilitate magnetization when an unmagnetized bar is between the poles of a magnet, because they assist the magnetizing force in breaking up the molecular groups or chains and in getting the molecules into alignment. Soft iron, then, has higher permeability than hard steel, merely because the molecules of the former substance do not offer so much resistance to a force tending to swing them into line as do those of the latter substance. Steel has, on the other hand, a much greater retentivity than soft iron, merely because its molecules are not so easily moved out of position when once they have been aligned.

12. Saturated Magnets. Strong evidence for the correctness of the above view is found in the fact that a piece of iron or steel cannot be magnetized beyond a certain limit, no matter how strong the magnetizing force. This limit probably corresponds to the condition in which the axes of all the molecules are brought into parallelism, Fig. 17. The magnet is then said to be *saturated,* since it is as strong as it is possible to make it.

Fig. 17. Supposed Condition of Molecules in Saturated Magnet

13. Earth's Magnetism. The fact that a compass needle always points north and south, or approximately so, indicates that the earth itself is a great magnet, having an S pole near the geographical north pole and an N pole near the geographical south pole; for the magnetic pole of the earth which is near the geographical north pole must of course be unlike the pole of a suspended magnet which points toward it, and the pole of the suspended magnet which points toward the north is the one which by convention it has been decided to call the north pole. The magnetic pole of the earth which is near the north geographical pole was found in 1831 by Sir James Ross in Boothia Felix, Canada, latitude 70° 30′ N., longitude 95° W. It was located again in 1905 by Captain Amundsen at a point a little farther west. Its approximate location is 70° 5′ N., and 96° 46′ W. It is probable that it slowly shifts its position.

14. Declination. It is, of course, on account of the fact that the earth's magnetic and geographical poles do not altogether coincide, that the magnetic needle does not point exactly north, and also that the direction in which it does point changes as the needle is moved about over the earth's surface. This last fact was first discovered by Columbus on his voyage to America, and caused great alarm among his sailors. There are other local causes, however, such as large deposits of iron ore, which cause local deviations of the needle from the true north. The number of degrees by which the needle varies from the north and south line at a given point, is called the *declination* at that point.

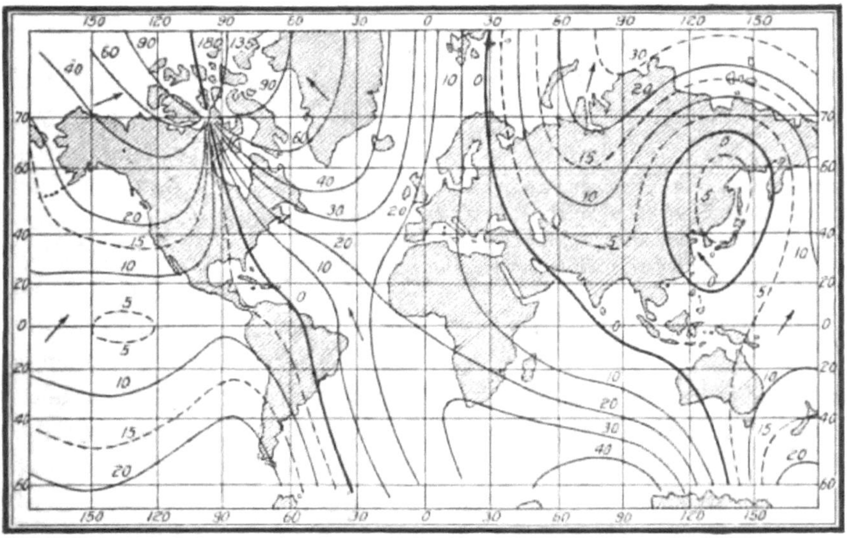

Fig. 18. Map Showing Lines of Equal Declination

Many of the early compass surveys made in the laying out of section lines, etc., are in error because the declination of the locality was not accurately known. Mariners also must have an accurate knowledge of the declination at all times in order to keep in their course. On this account, the United States Coast and Geodetic Survey in this country and similar departments of the governments of other countries spend thousands of dollars annually in collecting the data for maps which show the declination of any locality. Such a map is shown in Fig. 18, and the declinometer used for this work is shown in Fig. 19.

15. Inclination or Dip. Let an unmagnetized knitting needle *a*, Fig. 20, be thrust through a cork, and let a second needle *b* be passed through the cork at right angles to *a*. Let the system be adjusted by means of wax or a pin *c*, until it is in neutral equilibrium about *b* as an axis, when *a* is pointing east and west. Then let *a*

Fig. 19. Standard Declinometer Used in Coast Survey Work
Courtesy of U. S. Coast & Geodetic Survey, Washington, D. C.

be strongly magnetized by stroking one end of it from the middle out with the *N* pole of a strong magnet, and the other end from the middle out with the *S* pole of the same magnet. When this improvised dipping needle is replaced on its supports and turned into a north-and-south line with its *N* pole toward the north, it will be found, in the north temperate zone, to dip so as to make an angle of from 60° to 75° with the horizontal. This shows that in the latitudes mentioned the earth's magnetic lines are not at all parallel to the earth's surface. The angle between these lines and the earth's surface is called the *dip*, or *inclination*, of the needle. At Washington

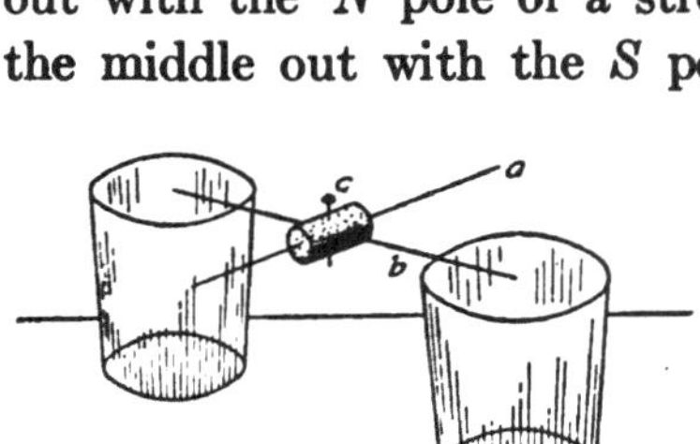

Fig. 20. Simple Experiment Illustrating Dip

it is 71° 5′; at Chicago, 72° 50′; at the magnetic poles it is of course 90°; and at the so-called *magnetic equator*—an irregular curved line passing through the tropics—the dip is 0°. The dipping needle used by the United States Coast and Geodetic Survey for these observations is shown in Fig. 21.

16. Earth's Inductive Action. A very instructive way of showing that the earth acts like a great magnet is to hold any ordinary iron or steel rod parallel to the earth's magnetic lines, that is, about in the geographical meridian, but with the north end slanting down at an angle of say 70°, and then to strike one end a few blows with the hammer. The rod will be found to have become a magnet with its upper end an *S* pole, like the north pole of the earth, and its lower end an *N* pole. If the rod is reversed and tapped again with the hammer, its magnetism will be reversed. If held in an east-and-west positon and tapped, it will become demagnetized, as is shown by the fact that both ends of it will attract either end of a compass needle

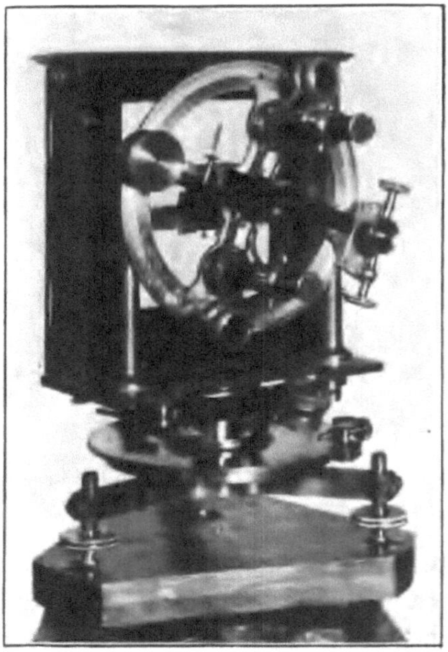

Fig. 21. Standard Dipping Needle Used in Coast Survey Work
Courtesy of U. S. Coast & Geodetic Survey, Washington, D. C.

EXAMPLES FOR PRACTICE

1. If a bar magnet is floated on a piece of cork, will it tend to float toward the north? Why?

2. Will a bar magnet pull a floating compass needle toward it? Compare the answer to this question with that to the preceding one.

3. Why should the needle, *a*, Fig. 20, be placed east and west, when adjusting for neutral equilibrium, before it is magnetized?

4. The dipping needle is suspended from one arm of a steel-free balance and carefully weighed. It is then magnetized. Will its apparent weight increase?

5. Does it matter whether a pocket compass is mounted in a brass case or an iron one?

6. When a piece of soft iron is made a temporary magnet by bringing it near the *N* pole of a bar magnet, will the end of the iron nearest the magnet be an *N* or an *S* pole?

7. Devise an experiment which will show that a piece of iron attracts a magnet just as truly as the magnet attracts the iron.

8. How would an ordinary compass needle act if placed over one of the earth's magnetic poles? How would a dipping needle act at this point?

9. With what force will an *N* magnetic pole of strength 6 attract, at a distance of 5 cm., an *S* pole of strength 1? Of strength 9? Ans. .24 dyne, 2.16 dynes

10. A magnetic *N* pole of the strength of 50 units is repelled by a second pole 10 cm. away with a force of 30 dynes. Find the strength of the second pole. Is it an *N* or an *S* pole? Ans. 60 units. *N* pole

STATIC ELECTRICITY

17. Electrification by Friction. Let a hard rubber (ebonite) rod be rubbed with flannel or cat's fur and then be brought near some dry pith balls or bits of paper, Fig. 22. These light bodies are found to be first attracted to the rod and then, after having been in contact with it for a very short time, to be repelled by it. These experiments may be easily tested at home by rubbing a hard rubber comb on the coat sleeve.

Fig. 22. Experiment Illustrating Electrification by Friction

This sort of attraction was observed by the Greeks as early as 600 B. C., when it was found that amber which had been rubbed with silk attracted various sorts of light bodies. It was not, however, until 1600 A. D. that Dr. William Gilbert, physician to Queen Elizabeth, and sometimes called the father of the modern science of electricity and magnetism, discovered that the effect could be produced by rubbing together a great variety of other substances besides amber and silk, such, for example, as glass and silk, sealing wax and flannel, and hard rubber and cat's fur.

18. Positive and Negative Electricity. Let a glass rod which has been electrified by rubbing it with silk be suspended by a silk thread, Fig. 23. Let an ebonite rod which has been rubbed with cat's fur be suspended in a second stirrup. If now a second glass rod which has been rubbed with silk is brought near the suspended glass rod, it will be found to repel it strongly; but when it is brought near the suspended ebonite, it will attract it no less strongly. On the other hand, a second electrified ebonite rod will attract the glass and repel the ebonite.

Evidently, then, the electrifications which have been imparted to the glass and to the ebonite are opposite, in the sense that an electrified body which attracts one repels the other. We say, therefore, that there are two kinds of electrification, and we arbitrarily call one *positive* and the other *negative*. Thus a *positively electrified body* is defined as one which acts with respect to other electrified bodies like *a glass rod which has been rubbed with silk*, and *a negatively electrified body* is one which acts like *an ebonite rod which has been rubbed with cat's fur.*

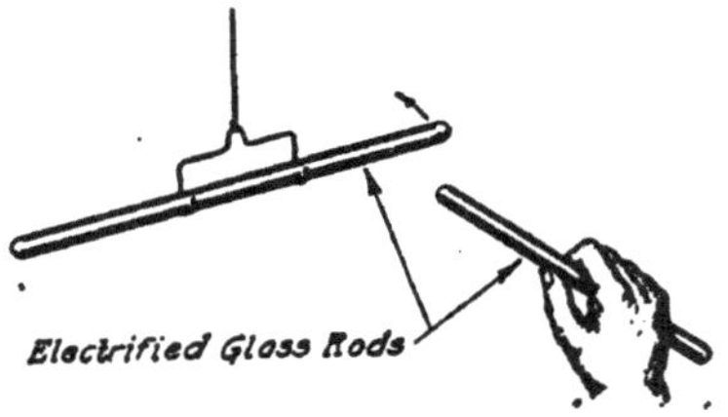

Fig. 23. Proof of Electrical Attraction and Repulsion

19. Laws of Electrical Attraction and Repulsion. The facts presented in the preceding experiment may be stated in the following law. *Electrical charges of like kind repel each other; those of unlike kind attract each other.* The forces of attraction or repulsion are found, like those of gravitation and of magnetism, *to decrease as the square of the distance increases.*

20. Measurements of Electrical Quantities. The fact of attraction and repulsion is taken as the basis for the definition and measurement of so-called *quantities* of electricity. Thus a small charged body is said to contain 1 electrostatic unit of electricity when it will repel an exactly equal and similar charge placed 1 centimeter away in air with a force of 1 dyne. The number of electrostatic units of electricity on any charged body is then measured by the number of dynes of force which it exerts upon a unit static charge placed at a given distance from it. For example, a charge which at a distance of 10 centimeters repels a unit charge with a force of 1 dyne contains 100 electrostatic units of electricity.

21. Coulomb's Law. The above laws were first quantitatively proved by Coulomb who used a torsion balance to measure the forces of attraction. He found that the force of attraction or repulsion, in dynes, between two charged bodies is given by the equation

$$F=\frac{Q_1 Q_2}{D^2} \quad \text{(Coulomb's Law)}$$

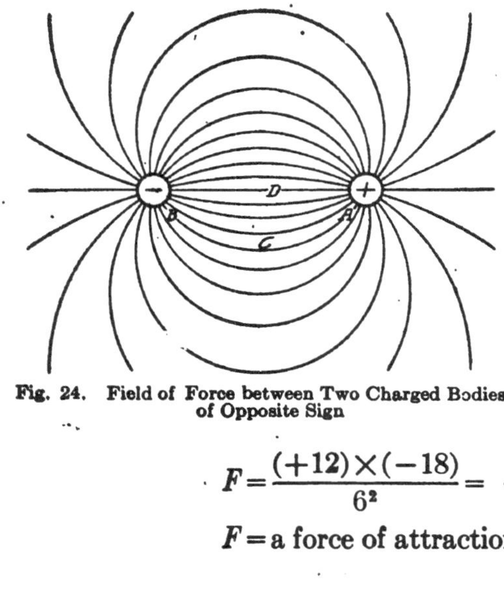

Fig. 24. Field of Force between Two Charged Bodies of Opposite Sign

where Q_1 and Q_2 represent the number of electrostatic units of electricity upon the two charged bodies, respectively, and D represents their distance apart in centimeters.

Example. With what force will a positive charge of 12 electrostatic units of electricity act upon a negative charge of 18 units when placed 6 cm. apart?

$$F=\frac{(+12)\times(-18)}{6^2}= -6$$

F = a force of attraction of 6 dynes

Ans. 6 dynes

Evidently the force is one of attraction whenever, as above, the result is minus, and one of repulsion whenever the result is plus, i.e., for like charges.

22. Electrical Field. If the two small bodies, A and B, Fig. 24, are given equal charges of positive and negative electricity, it is evident that a third small body with a positive charge when placed in the region about A and B would be repelled by A and attracted by B. Any line, such as ACB, Fig. 24, along which a plus charge would move in going from A to B is called an electrical line of force. The electrical field is the whole region about A and B in which a unit plus test charge would indicate any force acting upon it. The strength of the electrical field is represented by drawing as many electrical lines of force through each square centimeter as a unit plus test charge would experience dynes of force when placed in that locality. When A and B are alike in sign, say positive, the electrical field takes the form shown in Fig. 25.

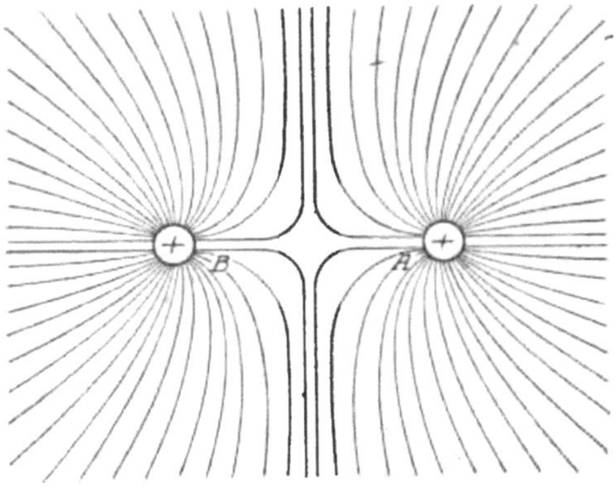

Fig. 25. Field of Force between Two Charged Bodies of the Same Sign

Example. How many ergs of work must be done against the electrical field, Fig. 24, in carrying a unit plus electrostatic charge from B to A, if the average force acting on this test charge is 2 dynes and the distance from B to A is 15 cm.? (An erg is a measure of the work done when a force of one dyne moves its point of application one centimeter.)

$$\begin{aligned} \text{work} &= \text{force} \times \text{distance} \\ &= 2 \times 15 \\ &= 30 \text{ ergs} \end{aligned}$$

Ans. 30 ergs

23. Conductors and Insulators. If a pith ball is in contact with a metal body A, Fig. 26, and if this body is connected by a wire to another metal body B, then, when B is rubbed with an electrified glass rod, A will be found immediately to repel the pith ball from itself. That is, a portion of the charge communicated to B evidently passes instantly over the wire to A. If the experiment is repeated when A and B are connected by a thread of silk, or by a rod of wood instead of metal, no effect will be observed at all upon the pith ball. If a moistened thread connects A and B, the pith ball will be affected, but not so soon as when A and B are connected by a wire.

Fig. 26. Experiment Illustrating Conductors and Nonconductors

These experiments make it clear that while electric charges pass with perfect readiness through a wire, they are quite unable to pass along dry silk or wool, while they pass with considerable difficulty along moist silk. We are therefore accustomed to divide substances into two classes, *conductors* and *nonconductors* or *insulators*, according to their ability to transmit electrical charges from point to point. Thus, metals and solutions of salts and acids in water are all conductors of electricity, while glass, porcelain, rubber, mica, shellac, wood, silk, vaseline, turpentine, paraffin, and oils generally are insulators. No hard and fast line, however, can be drawn between conductors and nonconductors, since all so-called insulators conduct to some extent, while the so-called conductors differ greatly among themselves in the facility with which they transmit charges.

The fact of conduction brings out sharply one of the most essential distinctions between electricity and magnetism. Magnetic poles exist only in iron and steel, while electrical charges can be communicated to any body whatsoever, provided they are insulated. These charges pass from point to point over conductors, and can be transferred by contact from one body to any other, while magnetic poles remain fixed in position, and are wholly uninfluenced by contact with other bodies, unless these bodies themselves are magnets.

24. Electrostatic Induction. If a metal ball A, Fig. 27, is strongly charged by rubbing it with a charged rod, and then brought near an insulated metal body B, which is provided with pith balls or strips of paper, a, b, c, as shown, the divergence of a and c, will show that the ends of B have received electrical charges because of the presence of A, while the failure of b to diverge will show that the middle of B is uncharged. Further, the rod which charged A will be found to repel c but to attract a. When A is removed, all evidences of electrification in B will disappear.

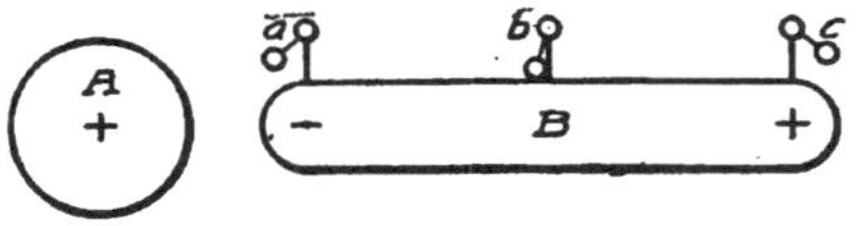

Fig. 27. Electrostatic Induction

From experiments of this sort, we conclude that when a conductor is brought near a charged body the end away from the charged body becomes electrified with the same kind of electricity as that on the charged body, while the end near the charged body receives a charge of opposite sign. *This method of producing elec-*

trification by the mere influence which an electric charge has upon a conductor placed in its neighborhood, is called electrostatic induction. The fact that as soon as A is removed, a and c collapse, shows that this form of electrification is only a temporary phenomenon.

25. Two-Fluid Theory of Electricity. We can describe the facts of induction conveniently by assuming that in every conductor there exists an equal number of positively and negatively charged corpuscles, which are very much smaller than atoms and which are able to move about freely among the molecules of the conductor. According to this view, when no electrified body is near the conductor B, it appears to have no charge at all, because all of the little positive charges within it counteract the effects upon outside bodies of all the little negative charges. But as soon as an electrical charge is brought near B, it drives as far away as possible the little corpuscles which carry charges of sign like its own, while it attracts the corpuscles of unlike sign. B, therefore, becomes electrified like A at its remote end, and unlike A at its near end. As soon as the inducing charge is removed, B immediately becomes neutral again because the little positive and negative corpuscles come together under the influence of their mutual attraction. This picture of the mechanism of electrification by induction is a modern modification of the so-called *two-fluid theory* of electricity, which conceived of all conductors as containing equal amounts of two weightless electrical fluids, called positive electricity and negative electricity. Although it is extremely doubtful whether this theory represents the actual conditions within a conductor, yet we are able to say with perfect positiveness that *the electrical behavior of a conductor is exactly what it would be if it did contain equal amounts of positive and negative electrical fluids*, or equal numbers of minute positive and negative corpuscles which are free to move about among the molecules of the conductor under the influence of outside electrical forces. Furthermore, since the real nature of electricity is as yet unknown, it has gradually become a universally recognized convention to speak of the positive electricity within a conductor as being repelled to the remote end, and the negative electricity as being attracted to the near end by an outside positive charge, and *vice versa*. This does not imply the acceptance of the two-fluid theory. It is merely a way of describing the fact that the remote end does acquire a charge like that of the inducing body, and the near end a charge unlike that of the inducing body.

26. Electron Theory. A slightly different theory is quite generally accepted by most physicists of high standing at the present time. According to this, a certain amount of positive electricity is supposed to constitute the nucleus of the atom of every substance. About this positive charge are grouped a number of very minute negatively charged corpuscles or electrons, the mass of each of which is approximately $\frac{1}{1760}$ of that of the hydrogen atom. The sum of the negative charges of these electrons is supposed to be just equal to the positive charge of the atom, so that in its normal condition, the whole atom is neutral and uncharged. But in the jostlings of the molecules of the conductor, electrons are continually getting loose from the atoms, moving about freely among the molecules, and re-entering other atoms which have lost their electrons. Therefore, at a given instant, there are always in every conductor a large number of free negative electrons and an exactly equal number of atoms which have lost electrons and which are therefore positively charged. Such a conductor would, as a whole, show no charge of either positive or negative electricity. But if a body charged, for example, negatively, were brought near such a body, the negatively charged electrons would stream away to the remote end, leaving behind them the positively charged atoms which are not supposed to be free to move from their positions. On the other hand, if a positively charged body is brought near the conductor, the negative electrons are attracted and the remote end is left with the immovable positive atoms.

The only advantage of this theory over that suggested in the preceding section, in which the existence of both positive and negative corpuscles is assumed, is that there is much direct experimental evidence for the existence of free negatively charged corpuscles of about $\frac{1}{1760}$ the mass of the hydrogen atom, but no direct evidence as yet for the existence of positively charged bodies smaller than atoms.

The charge of one electron is called the elementary electrical charge. Its value has recently (1913) been accurately measured, and found to be very nearly one two-billionth part of the electrostatic unit of charge defined in section 20. Every electrical charge consists of an exact number of these elementary electrical charges (atoms of electricity) scattered over the surface of the charged body.

27. Gold-Leaf Electroscope. One of the most sensitive and convenient instruments for detecting the presence of an electrical charge upon a body and for determining the sign of that charge, is the *gold-leaf electroscope*, Fig. 28. It consists of a glass jar, through the neck of which passes a metal rod supported by a rubber stopper or some other insulated material, and carrying at its lower end two gold leaves or strips of aluminum foil. To detect with this instrument the *presence* of an electrical charge, it is only necessary to bring near the upper end of the electroscope the body which is to be tested. If it is charged, it will repel electricity of the kind which it possesses to the leaves, and draw the unlike kind to the upper end. The leaves under the influence of the like charges which they possess will stand apart or diverge. If the body is not charged, the gold leaves will not be affected at all.

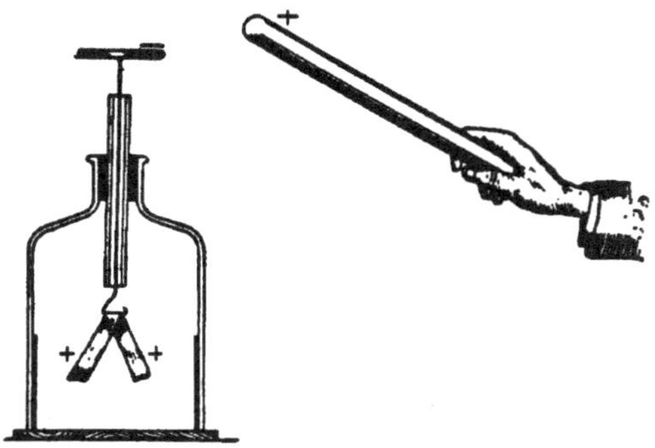

Fig. 28. Gold-Leaf Electroscope

To determine the *sign* of an unknown charge with an electroscope, we first impart a charge of known sign to the electroscope by touching it, for example, with a piece of sealing wax, which has been rubbed with cat's fur. This charges the leaves negatively and causes them to diverge. The unknown charge is then slowly brought near the upper end of the electroscope; and if the divergence of the leaves is *increased*, the sign of the unknown charge is negative, for the increased divergence means that more negative electricity has been repelled to the leaves. If the divergence is *decreased* instead of increased, the sign of the unknown charge is *positive*, for the decreased divergence of the leaves means that a part of the negative electricity already on the leaves has been drawn to the upper end.

28. Charging by Induction. If a positively charged body C, Fig. 29, is brought near two conductors A and B in contact, we have seen that a positive charge will appear upon A and a negative charge upon B. If C were removed these charges would recombine and A and B both become neutral. But if, before C has been removed, A and B are separated, and if then C is removed, there is no opportunity for this recombination. Hence A is left permanently charged positively, and B negatively. These charges can be easily detected by bringing A and B into the neighborhood of a charged electroscope. One will cause the divergence of the leaves to increase, the other will cause it to decrease.

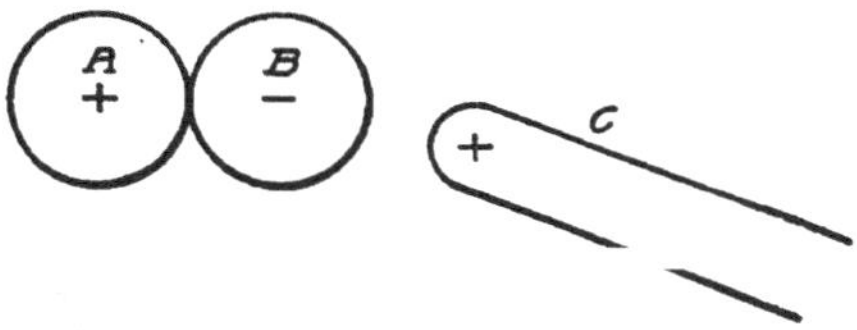

Fig. 29. Charging Two Conductors by Induction

Again, if a positively charged body C, Fig. 30, is brought near the conductor B, and if, while C is still in position, the finger is touched anywhere to the conductor B and then removed, then, when C is removed, B is found to be negatively charged. In this case the body of the experimenter corresponds to the conductor A of the preceding experiment, and removing the finger from B corresponds therefore to separating the two conductors A and B. In the use of this method of charging a single body by induction, it makes no difference with the sign of the charge left upon B where the finger touches the body B, whether at a or at b or at any other point, for it is always the kind of electricity which is like that on the charging body C that is repelled off to earth through the finger; while the charge which is unlike that upon C is drawn to the part of B which is next to C, and as soon as C is removed this spreads over the whole body B. Whenever, then, a single body is charged by induction in this manner, *the sign of the charge left upon it is always opposite to that of the inducing charge.*

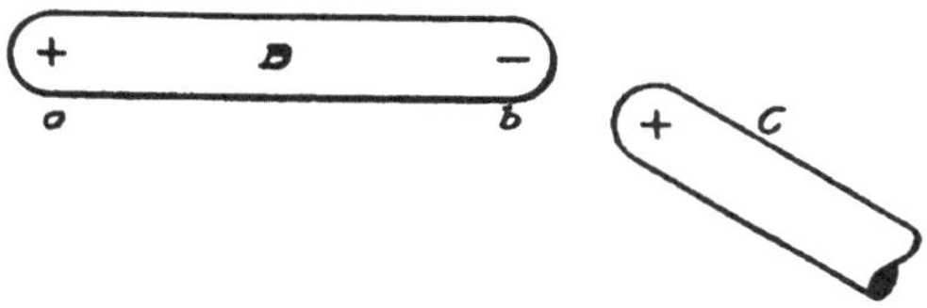

Fig. 30. Charging a Single Conductor by Induction

Thus, if we wished to charge the electroscope, Fig. 31, negatively by induction from a positively charged glass rod, we should first bring the rod near the knob of the electroscope, tnus causing the leaves to diverge because of the positive electricity which is repelled to them. Then while the rod is still in position near the electroscope, if we should touch the knob of the latter with the finger the leaves would at once collapse. This is because the positive electricity on the electroscope passes off to earth through the finger, while the negative is held attracted to the knob of the electroscope by the positive charge on the rod. In this condition the negative is sometimes said to be *bound* by the attraction of the positive charge on *C*. We should then remove the finger and finally the rod. The negative would then be free to pass to the leaves and cause them to diverge. The electroscope would thus be charged negatively. This is often one of the most convenient methods of charging an electroscope. It should always be used where it is desired to obtain a charge of sign opposite to that of the charging body. If it is desired to obtain a charge on the electroscope of sign like to that of the charging body we simply touch the body directly to the knob of the electroscope, and thus charge it by conduction rather than by induction.

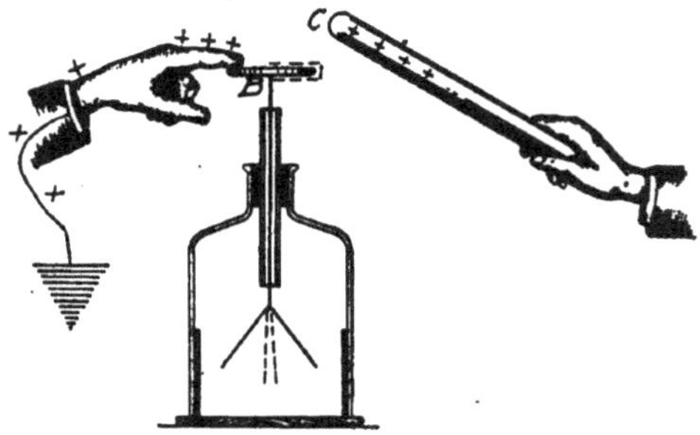

Fig. 31. Charging an Electroscope by Induction

One advantage of charging by induction lies in the fact that the charging body loses none of its own charge, whereas, in charging by conduction, the charging body must of course part with a portion of its charge.

29. Appearance of Positive and Negative Electricities Simultaneous and in Equal Amounts. If a strip of flannel is stuck fast to one side of a rod of sealing wax and rubbed back and forth over a second rod of sealing wax, and if then the two bodies are brought near the knob of a charged electroscope before they are separated, it is found that they give no evidence at all of electrification. But if they are separated and brought in succession to the knob of the electroscope, they will exhibit positive and negative charges of equal strength, the flannel being positive, and the bare sealing wax negative. Similarly, when a glass rod is charged positively by rubbing it with silk, the silk when tested is always found to possess a negative charge. These experiments show that in producing electrification by friction, positive and negative charges appear simultaneously and in equal amount. This confirms the view, already brought forward in connection with induction, that the process of electrification always consists in a separation of positive and negative charges which already exist in equal amounts within the bodies in which the electrification is developed. Certain it is that it is never possible to produce in any way whatever, one kind of electricity without producing at the same time an equal amount of the opposite kind.

30. Electrical Charge Resides upon Outside Surface of Conductor. If a deep metal cup is placed upon an insulating stand and charged as strongly as possible, either from a charged rod or from an electrical machine, Fig. 32, and a metal ball suspended by a silk thread is touched to the inside of the cup, the ball is found upon removal to show no evidence of charge when brought near the knob of the electroscope. If, on the other hand, the ball is touched to the outside of the cup, it exhibits a strong charge. Or, again, if the metal ball is first charged and then touched to the inside of the cup, it loses its charge completely, even though the cup itself may be very strongly charged. These experiments show that an electrical charge resides entirely on the outside surface of a conductor. This is a result which might have been inferred from the fact that all the little electrical charges of which the total charge is made up repel each other and therefore move through the conductor until they are on the average as far apart as possible, that is, until they are all upon the surface.

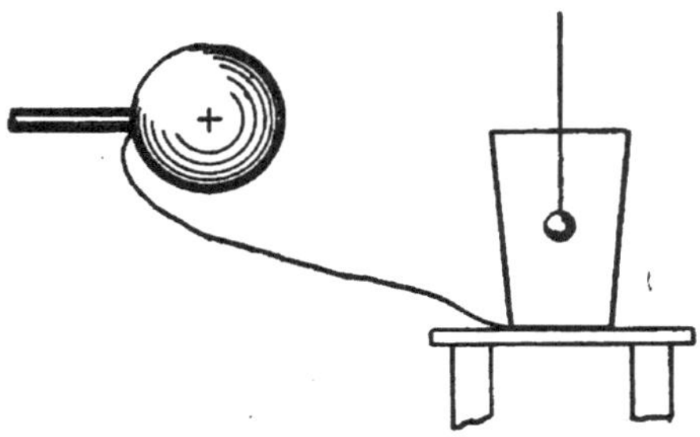

Fig. 32. Proof that Electric Charge is on Outside of Conductor

31. Greatest Density of Charge Where Curvature of Surface is Greatest. Since all of the parts of an electrical charge tend, because of their mutual repulsions, to get as far apart as possible, we might infer that if a charge of either sign is placed upon an oblong conductor, like that of Fig. 33, *a*, it would distribute itself so that the electrification at the ends would be stronger than that at the middle. The correctness of this inference is easy to verify experimentally, for it is only necessary to attach a penny to the end of a piece of sealing wax and touch it first to the middle of a long charged conductor, next to bring it over the knob of the electroscope; and then to repeat the operation when the penny is touched to the end of the conductor. The electroscope will be affected much more strongly in the latter case than in the former. If we should test in this way the distribution on a pear-shaped body, Fig. 31, *b*, we should find the density of electrification considerably greater on the small end than on the large end. By density of electrification is meant the quantity of electricity on a unit-area of the surface.

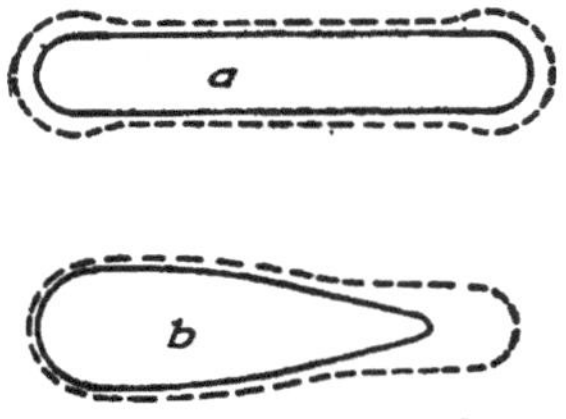

Fig. 33. Distribution of Charge on Different Shaped Conductors

32. Discharging Effect of Points. It might be inferred from the above that if one end of a pear-shaped body is made more and more pointed, then, when a charge is imparted to the body, the electric density on the small end will become greater and greater as the curvature of this end is made sharper and sharper. That this is the case is indicated by the effect which experiment shows that points have upon electrical charges; for if a very sharp needle is attached to any insulated conductor which is provided with paper or pith ball indicators (see *B* of Fig. 27), it is found impossible to impart to *B* a permanent charge; that is, if one attempts to charge it by rubbing over it a charged glass rod or other charged body, the indicators will be found to collapse as soon as the rod is removed. That this is due to an effect of the point can be proved either by removing the needle, or by covering up the point with wax, when the charge will be retained, as in the case of any insulated body. The probable explanation of the phenomenon is as follows:

The density of the charge becomes so intense upon the point that the molecules of air immediately adjoining the point are broken apart into positive and negative parts, and portions which are of unlike sign to the charge on the point are attracted to it, thus neutralizing the charge upon the body, while portions of like sign are repelled.

The effect of points upon an electrical charge may be strikingly shown by holding a very sharp needle in the hand and bringing it toward the knob of a charged electroscope. The leaves will fall together rapidly. Or, if the needle is brought near a tassel of tissue paper which is attached to an electrified conductor, Fig. 34, the electrified streamers, which stand out in all directions because of their mutual repulsions, will at once fall together. In both of these cases, the needle becomes electrified by *induction* and discharges to the knob of the electroscope or to the tassel electricity of the opposite sign to that which it contains, thus neutralizing its charge.

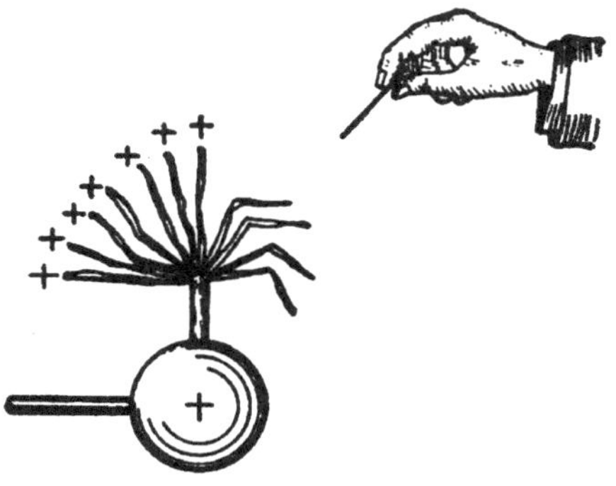
Fig. 34. Electrified Tassel Discharged by Needle Point

An interesting variation of the last experiment is to mount an electric whirl, Fig. 35, upon one knob of an electrical machine. As soon as the machine is started, the whirl will rotate rapidly in the direction of the arrow. The explanation is as follows: On account of the great magnitude of the electric force near the points, the molecules of the gas just in front of them are broken into positive and negative parts. The part of sign unlike that of the charge on the points is drawn to them, while the other part is repelled. But since this repulsion is mutual, the point is pushed back with the same force with which the particles are pushed forward; hence the rotation. The repelled particles in their turn drag the air with them in their

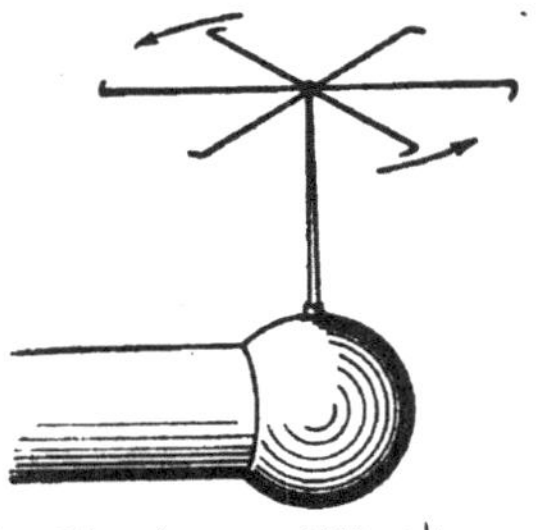
Fig. 35. Action of Electric Whirl

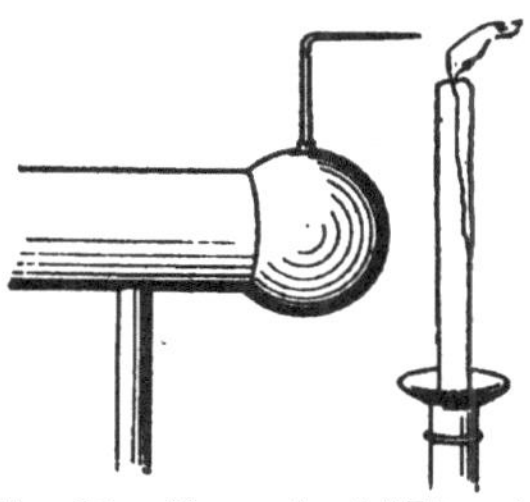
Fig. 36. Example of "Electric Wind"

forward motion, and thus produce the *electric wind*, which may be detected easily by the hand or by a candle held in front of the point, Fig. 36.

33. Lightning and Lightning Rods. It was in 1752 that Franklin, during a thunderstorm, sent up his historic kite. This kite was provided at the top with a pointed wire. As soon as the hempen kite string had become wet, he succeeded in drawing ordinary electric sparks from a key attached to the lower end. This experiment demonstrated for the first time that thunderclouds carry ordinary electrical charges which may be drawn from them by points, just as the charge was drawn from the tassel in the experiment of Section 32. It also showed that lightning is nothing but a huge electric spark. Franklin applied this discovery in the invention of the lightning rod. The way in which the rod discharges the cloud and protects the building is as follows: As the charged cloud approaches the building, it induces an opposite charge in the rod. This induced charge escapes rapidly and quietly from the sharp point in the manner explained above and thus neutralizes the charge of the cloud.

Example. Let a metal plate C, Fig. 37, be supported above a metal ball E, and let C and E be attached to the two knobs of an electrical machine. When the machine is started, sparks will pass from C to E, but if a point p is connected to E, the sparking will cease; that is, the point will protect E from the discharges, even though the distance Cp be considerably greater than CE.

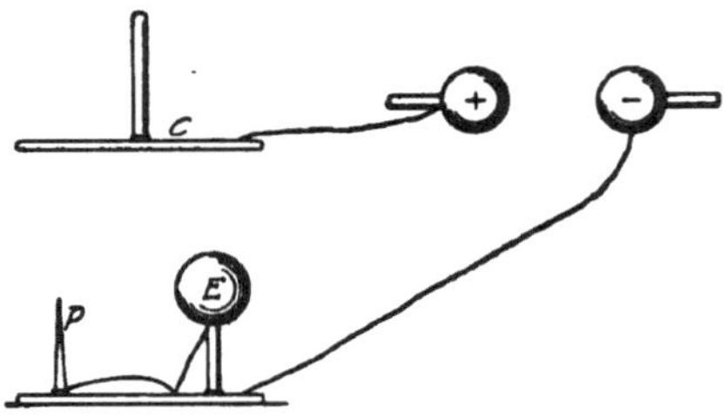

Fig. 37. Experiment Illustrating Action of Lightning Rod

The lower end of a lightning rod should be buried deep enough so that it will always be surrounded by moist earth, since dry earth is a poor conductor. It will be seen, therefore, that lightning rods protect buildings not because they conduct the lightning to earth, but because they prevent the formation of powerful charges in the neighborhood of buildings on which they are placed.

34. Smoke and Fume Condenser. An interesting application of the discharge from points can be made in connection with the precipitation of smoke and injurious fumes from factories, smelters, etc. If in Fig. 37, the plate C and the point p were placed over an aperture through which fumes or smoke were passing so that the fumes would have to pass between the point and the plate, and if, at the same time, by means of some source of electricity the point and plate were kept oppositely charged, a peculiar effect would be noticed. As the discharge from the point p was continuous, the effect would be to strongly charge all particles in the space with the same kind of electricity as that coming from the point p. All of these particles so charged would immediately move toward the plate C and would condense upon it, thereby removing these particles from the air or gas passing out of the aperture.

A commercial application of this principle has been made in a great many smelters and other factories where obnoxious or injurious fumes are the result of some chemical or other operation in the factories. By suitably equipping the chimney or other passageway through which the fumes are carried with a plate and with discharge points at a proper distance away from this plate, and heavily charging the apparatus with a high electrical potential, very effective precipitation of these injurious fumes and smoke can be brought about. The potential must be very high, this being brought about, as will be learned later in the study of electricity, by using an alternating current of the usual voltage and transforming this current to a voltage of 20,000 to 75,000. By a suitable revolving commutator the current is made into a high potential direct current which charges the points and the plate. The discharge points are obtained by using asbestos or mica wool held between twisted strands of wire which carry the current, the many streamers of asbestos wool or the fine edges of the mica providing the great number of discharge points necessary. This apparatus has been found very effective in most cases but will precipitate only material particles. In other words, if the temperature of the gases emitted is so high that certain objectionable solids are in the vapor state (as arsenic, for example), this dangerous fume will pass through without being precipitated. By providing a second device after the fumes have cooled, the arsenic will be condensed. The device can easily be applied to the precipitation of coal smoke but some method must be provided for removing the particles of soot which gather upon the charged plate and interfere with the proper working of the apparatus.

35. Electrical Screens. We have seen that if a positively charged body A, Fig. 38, is brought near an uncharged conductor B, negative electricity is attracted to the near end of the conductor and positive electricity appears at the remote end, this separation of the positive and negative charges being dependent upon the presence of the charged conductor A. Let us see how this known fact as to the condition of the charges may be used to determine the electrical condition at any point p within the conductor. Since the electricity within the conductor is free to move under the influence of any electrical force which is acting upon it, it is clear that the accumulation of nega-

Fig. 38. Electric Condition at any Point Inside a Conductor

tive electricity at one end and of positive at the other will cease only when all electrical forces inside the conductor are reduced to zero, that is, when the charges on A and B, acting jointly, neutralize one another completely at any point p within the conductor. It appears, therefore, that the distribution of the induced charge on the surface of a conductor in the electrical field of a charged body must always be such that there is no force whatever inside the body. This theoretical conclusion was first experimentally verified by Faraday, who coated a large box with tinfoil and went inside the box with delicate electroscopes. He found that these electroscopes showed no effects whatever, even when powerfully charged bodies were brought near the outside of the box. The experiment is often repeated in a small way by placing an electroscope under a wire cage of rather small mesh, Fig. 39. A charged rod brought near the cage will produce no effect whatever upon the electroscope. We thus learn that electrical influences can be completely cut off from a body by surrounding it on all sides with a conductor.

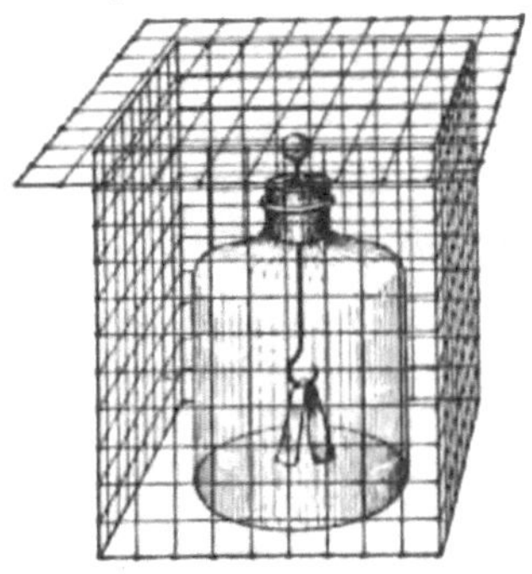

Fig. 39. Use of Electric Screens

36. Electrical Potential. There is a very instructive analogy between the use of the word *potential* in electricity and *pressure* in hydrostatics. For example, if water will flow from tank A, Fig. 40, to tank B through the connecting pipe R, we infer that the hydrostatic pressure at a must be greater than that at b, and we attribute the flow directly to this difference in pressure. In precisely the same way, if, when two bodies A and B, Fig. 41, are connected by a conducting wire r, a charge of positive electricity is found to pass from A to B, or of negative from B to A, we are accustomed to say that the electrical potential is higher at A than at B, and we assign this *difference of potential* as the cause of the flow. Thus, just as water tends to flow from

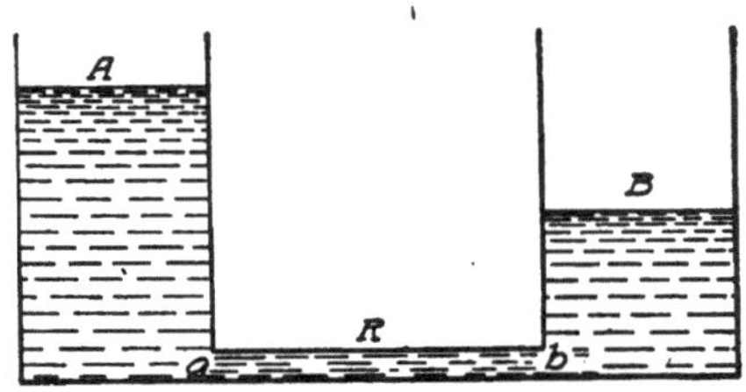

Fig. 40. Water Analogy Illustrating Electric Potential

points of higher hydrostatic pressure to points of lower hydrostatic pressure, so electricity is conceived of as tending to flow from points of higher electrical pressure or potential to points of lower electrical pressure or potential.

Again, if water is not continuously supplied to one of the tanks, Fig. 40, we know that the pressures at *a* and *b* must soon become the same. Similarly, if no electricity is supplied to the bodies *A* and *B*, Fig. 41, their potentials very quickly become the same. In other words, *all points on a system of connected conductors in which the electricity is in a stationary, or static, condition are necessarily at the same potential*; for if this were not the case, then the electricity which we imagine all conductors to contain would move through the conductor until the potentials of all points were equalized. In other words, equality in the potentials of all points on a conductor in the static condition follows at once from the fact of mobility of electrical charges through or over a conductor.

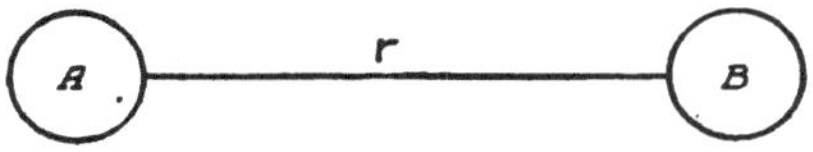

Fig. 41. Equality of Potential between Connected Bodies

But if water is continually poured into *A*, Fig. 40, and removed from *B*, the pressure at *a* will remain permanently above the pressure at *b*, and a continuous flow of water will take place through *R*. Similarly if *A*, Fig. 41, is connected with an electrical machine and *B* to earth, a permanent potential difference will exist between *A* and *B* and a continuous current of electricity will flow through *r*. Difference in potential is commonly denoted simply by the letters p.d. (potential difference).

When we speak simply of the *potential* of a body we mean the *difference of potential* which exists between the body and the earth, for the electrical condition of the earth is always taken as the zero to which the electrical conditions of all other bodies are referred. Thus a body which is positively charged is regarded as one which has a potential higher than that of the earth, while a body which is negatively charged is looked upon as one which has a potential lower than that of the earth. Fig. 42 represents the hydrostatic analogy of positively and negatively charged bodies. Since it has been decided to regard the flow of electricity as taking place from a point of higher potential to one of lower, it will be seen that when a discharge takes place between a negatively charged body and the earth, we must regard the positive electricity as passing from the earth to the body, rather than the negative as passing from the body to the earth. This is, indeed, a mere convention, but it is one which it is very important to remember in connection with the study of current electricity. From the point of view of the electron theory, Section 26, it would be natural to invert this convention exactly, since this theory regards the negative electricity alone as moving through conductors. But since the opposite convention has become established, it will not be wise to attempt to change it until the electron theory has become more thoroughly established than is at present the case.

37. Unit of P.D. The difference in the pressures produced by the water in tank *A* and in tank *B*, in grams per square centimeter, Fig. 40, is numerically equal to the work done in gram-centimeters, in carrying 1 cubic centimeter of water from the level of the water in tank *B* up to the level of the water in tank *A*. This is true if any liquid other than water be taken. Thus the work done in carrying unit volume of the liquid from one level to the other against the force of gravity may be taken as a measure of the difference in pressure produced by the two columns of liquid. Similarly in Fig. 24, the difference in electrical pressure between *A* and *B* may be measured by the amount of work done when a unit plus charge is carried against the electrical field from *B* to *A*. The number of absolute units of p.d. between *A* and *B* is then taken to be equal to the number of ergs of work required to carry the unit plus charge from *B* to *A*.

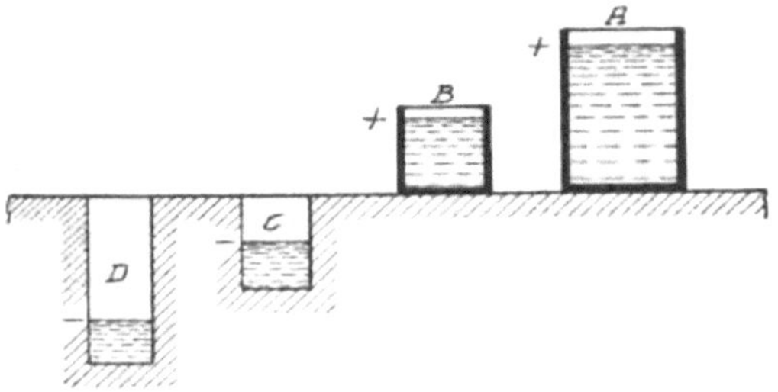

Fig. 42 Water Analogy of High and Low Potentials

The absolute unit of p.d. is too large for practical purposes and $\frac{1}{300}$ as large a unit has been chosen for the practical unit of p. d. and is called the *volt* in honor of the famous Italian physicist, Allessandro Volta (1745-1827), who about 1800, at the University of Pavia, first made quantitative measurements upon the potentials of charged bodies.

A volt is then defined as the p. d. between two charged bodies when it requires $\frac{1}{300}$ of an erg of work to carry a unit plus electrostatic charge from one body to the other against the electrical field existing between the two bodies.

The volt is approximately equal to the electrical pressure between the ends of a strip of copper and a strip of zinc immersed in dilute sulphuric acid. (See Fig. 53.)

Examples. 1. If in the problem given at the end of Section 22, it required 30 ergs of work to carry a unit plus electrostatic charge from *B* to *A*, what was the p.d. between *A* and *B*, first, in absolute units of p.d., and second, in volts?

Since the potential difference in absolute units is numerically equal to the work done in ergs in carrying a unit plus charge from *B* to *A*, therefore the p.d. is 30 absolute units of p.d. The volt being only $\frac{1}{300}$ as large as the absolute unit there will be 300 times as many units to express the same quantity or

$$300 \times 30 = 9{,}000 \text{ volts} \qquad \text{Ans. } 9{,}000 \text{ volts}$$

2. The quantity of electricity which passes through the filament of a certain 16-candle-power tungsten (Mazda) lamp is 540,000,000 electrostatic units per second. How many ergs of work are done per second in forcing this electricity through the lamp if the p.d. across the lamp is 112 volts? What is the rating of the lamp in watts?

The number of ergs of work done per second i

$$\frac{540{,}000{,}000 \times 112}{300} = 201{,}600{,}000$$

Since 10^7 ergs per second, or 1 joule per second, is 1 watt, the rating of the lamp is

$$201{,}600{,}000 \div 10^7 = 20.16 \text{ watts} \qquad \text{Ans. } 20.16 \text{ watts}$$

NOTE: A 16-candle-power Mazda lamp is rated at 20 watts by the makers.

38. Some Methods of Measuring Potentials. One of the simplest methods of comparing the potential difference which exists between any two charged bodies and the earth, is to connect the charged bodies successively to the knob of an electroscope which has an outside case of metal or metal strips, as shown in Fig. 28, connected electrically to earth. The amount of separation of the gold leaves is then a measure of the p.d. between the earth and the charged body. If the electroscope is calibrated in volts, its reading gives the p.d. between the body and the earth. Such calibrated electroscopes are called *electrostatic voltmeters.* They are the simplest and in many respects the most satisfactory forms of voltmeters to be had. Their use, both in laboratories and in electrical power plants, is rapidly increasing. They can be made to measure a p.d. as small as $\frac{1}{1000}$ volt and as large as 200,000 volts. Fig. 43 shows one of the simpler forms of the electrostatic voltmeter. The outer case is of metal and is connected to earth at the point *a*. The body whose potential is sought is connected to the knob *b*. This is in metallic contact with the light aluminum vane *c*, which takes the place of the gold leaf.

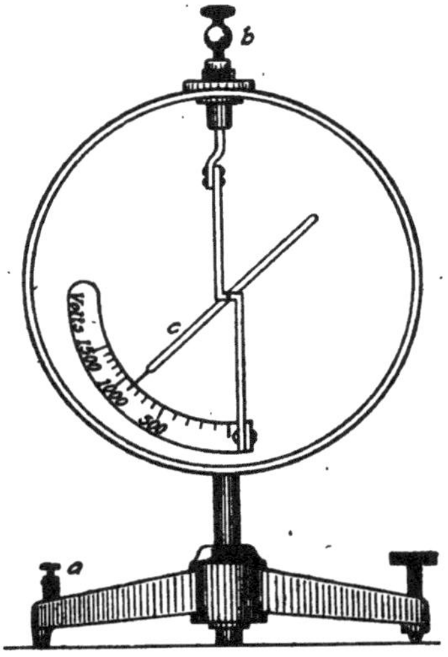

Fig. 43. Simple Electrostatic Voltmeter

A very convenient way of measuring a *large* p.d. without a voltmeter is to measure the length of the spark which will pass between the two bodies whose p.d. is sought. The p.d. is roughly proportional to spark length, each centimeter of spark length representing a p.d. of about 30,000 volts if the electrodes are large compared to their distance apart.

39. Condensers. If a metal plate *A* is mounted on an insulating plate and connected with an electroscope, Fig. 44, and if a

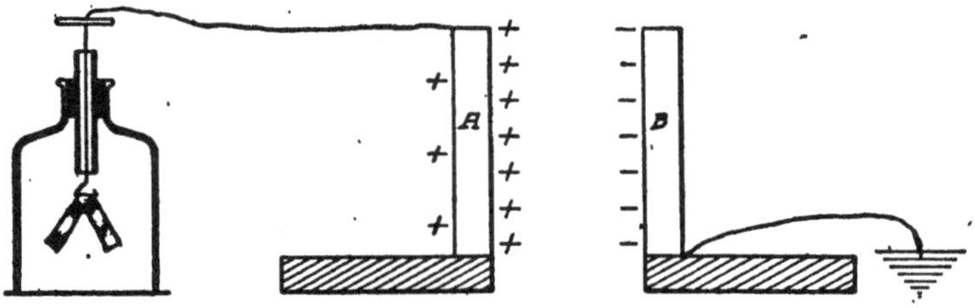

Fig. 44. Diagram Illustrating Action of a Condenser

second plate *B* is similarly mounted and connected to earth, then, when a charge is placed on *A*, it will be found that the gold leaves fall together as B approaches A and diverge farther as B recedes from A. This shows that the potential of A is diminished by bringing B close to it, in spite of the fact that the quantity of electricity on A has remained unchanged. If we convey additional plus charges to A, we find that many times the original amount of electricity may be placed on A when B is close to it, before the leaves return to their original divergence, that is, before the body regains its original potential.

We say, therefore, that the *capacity* of A for holding electricity has been very greatly increased by bringing near it another conductor which is connected to earth. It is evident from this statement that *we measure the capacity of a body by the amount of the electricity which must be put upon it in order to raise its potential to a given point.* The explanation of the increase of capacity in this case is obvious. As soon as B was brought near to A, it became charged, by induction, with electricity of sign opposite to that of A, the electricity of sign like that of A being driven off to earth through the connecting wire. The attraction between these opposite charges on A and B drew the electricity on A to the face nearest to B, and removed it from the more remote parts of A, so that it became possible to put a very much larger charge on A before the tendency of the electricity on A to pass over to the electroscope became as great as at first, that is, before the potential of A rose to its original value. Under circumstances of this sort the electricity on A is said to be *bound* by the opposite electricity on B.

An arrangement of this sort consisting of two conductors separated by a nonconductor is called a *condenser*. If the conducting points are very close together and one of them is joined to earth, the capacity of the system may be thousands of times as great as that of one of the plates alone.

40. Leyden Jar. The most common form of condenser is a glass jar coated with tinfoil part way to the top inside and outside, Fig. 45. The inside coating is connected by a chain to the knob, while the outside coating is connected to earth. Condensers of this sort first came into use in Leyden, Holland, in 1745. Hence they have since been called *Leyden jars*.

Such a jar is charged by holding the knob in contact with one terminal of an electrical machine and connecting the outer coat to earth either by a wire or simply by holding it with the hand. As fast as electricity passes to the knob, it spreads to the inner coat of the jar, where it attracts electricity of the opposite kind from the earth to the outer coat, repelling electricity of the same kind. If the inner and outer coatings are now connected by a discharging rod, Fig. 45, a very powerful spark will be produced. If a charged jar is placed on a glass plate so as to insulate the outer coat, the knob may be touched with the finger and no appreciable discharge noticed. Similarly the outer coat may be touched with the finger with the same result. But if the inner and outer coats are connected with the discharge, a powerful spark passes.

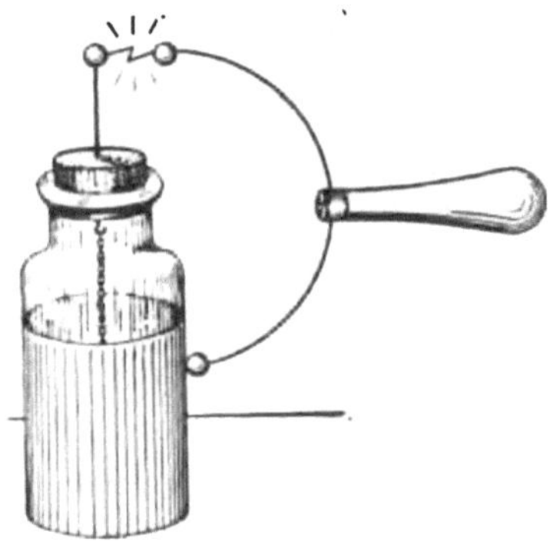

Fig 45. Discharge of a Leyden Jar

The experiment shows that it is impossible to discharge one side of the jar only, for practically all of the charge is *bound* by the opposite charge on the other coat. Therefore, the full discharge can occur only when the inner and outer coats are connected.

ELECTRICAL GENERATORS

41. Electrophorus. The electrophorus is a simple electrical generator which illustrates well the principle underlying the action of all electrostatic machines. All such machines generate electricity primarily by induction, not by friction. *B*, Fig. 46, is a hard rubber plate which is first charged by rubbing it with fur or flannel. *A* is a metal plate provided with an insulating handle. When the plate *A* is placed upon *B*, touched with the finger, and then removed, it is found possible to draw a spark from it, which in dry weather may be a quarter of an inch or more in length. The process may be repeated an indefinite number of times without producing any diminution in the size of the spark which may be drawn from *A*.

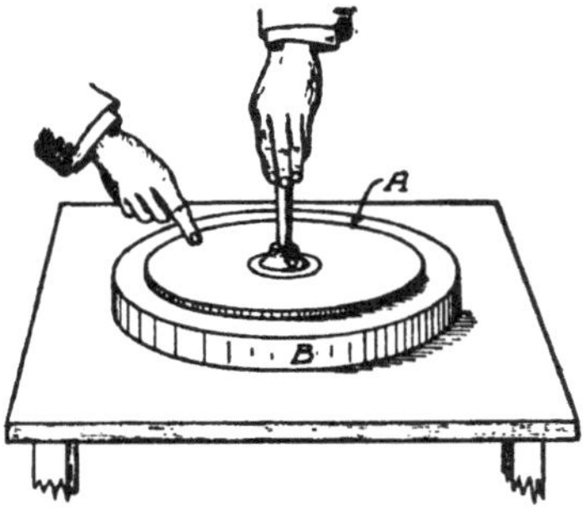

Fig. 46. Charging an Electrophorus

If the sign of the charge on *A* is tested by means of an electroscope, it will be found to be positive. This proves that *A* has been charged by induction, not by contact with *B*, for it is to be remembered that the latter is charged negatively. The reason for this is that even when *A* rests upon *B* it is in reality separated from it, at all but a very few points, by an insulating layer of air; and, since *B* is a nonconductor, its charge cannot pass off appreciably through these few points of contact. It simply repels negative electricity to the top side of the metal plate *A*, and draws positive to the lower side of this plate. The negative passes off to earth when the plate is touched with the finger. Hence, when the finger is removed and *A* lifted, it possesses a strong positive charge.

42. Source of the Energy of the Charge Obtained from an Electrophorus. Although an indefinite amount of electricity may be developed by an electrophorus and stored up, for example, in a Leyden jar, without diminishing at all the charge on *B*, yet this electrical energy does not come into existence without the expenditure of a corresponding amount of mechanical work; for each time that the plate *A* is removed from *B* it must be lifted against the attractions of the opposite charges on *A* and *B*, Fig. 47.

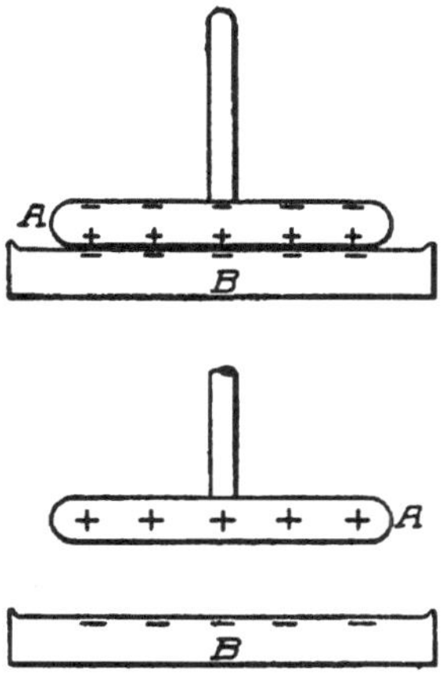

Fig. 47. Action on Conductor when Charged by Electrophorus

According to the principle of the conservation of energy, the energy in one charge on the plate *A* is then exactly equal to the work done against these electrical attractions in lifting the plate. This electrical energy is transformed into heat energy in the spark which is produced by the discharge.

If a bunsen burner is held in the hand, and a spark allowed to pass off the edge of *A* to the top of the burner while the gas is turned on, the gas will be instantly ignited by the spark.

43. Toepler-Holtz Static Machine. The ordinary static machine is nothing but a continuously acting electrophorus. Fig. 48 represents one type of such machine. Upon the back of the stationary glass plate *E* are pasted paper sectors, beneath which are strips of tinfoil *AB* and *CD*, called *inductors*. In front of *E* is a revolving glass plate carrying disks *l*, *m*, *n*, *o*, *p*, and *q*, called

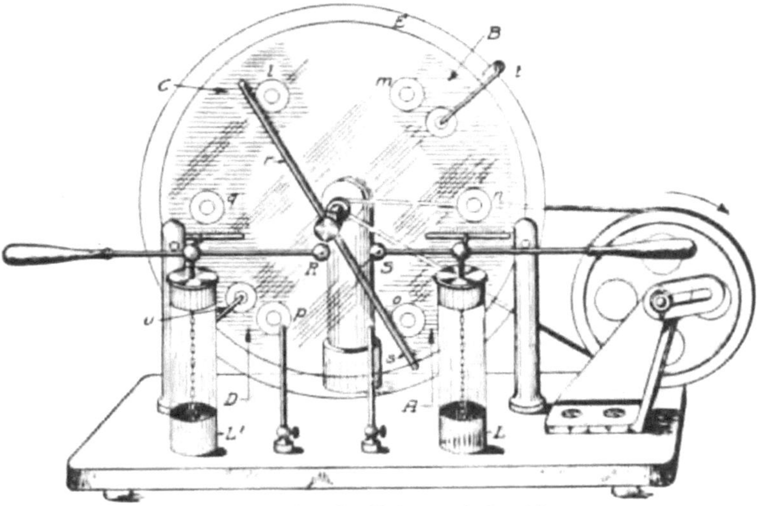

Fig. 48. Toepler-Holtz Static Machine

carriers. To the inductors *AB* and *CD* are fastened metal arms *t* and *u*, which bring these inductors into electrical contact with disks *l*, *m*, *n*, *o*, *p*, and *q*, when these disks pass beneath the tinsel brushes carried by *t* and *u*. A stationary metallic rod *rs* carries at its ends stationary brushes as well as sharp-pointed metallic combs. The two knobs *R* and *S* have their capacity increased by the Leyden jars *L'* and *L*.

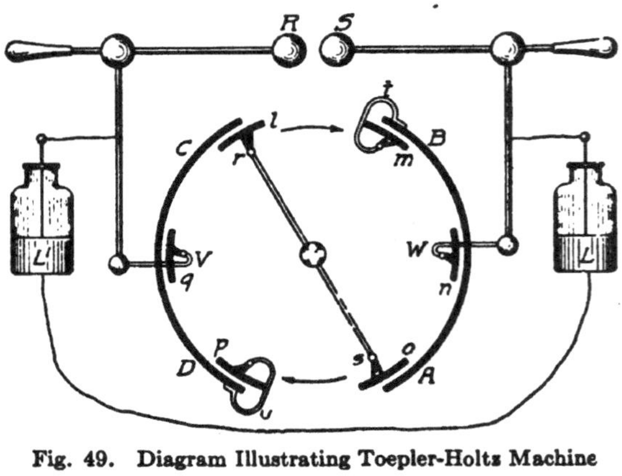

Fig. 49. Diagram Illustrating Toepler-Holtz Machine

The action of the machine is best understood from the diagram, Fig. 49. Suppose that a small + charge is originally placed on the inductor *CD*. Induction takes place in the metallic system consisting of the disks *l* and *o* and the rod *rs*, *l* becoming negatively charged, and *o* positively charged. As the plate carrying *l*, *m*, *n*, *o*, *p*, *q* rotates in the direction of the arrows, the negative charge on *l* is carried over to the position *m*, where a part of it passes over to the inductor AB, thus charging it negatively. When l reaches the position n, the remainder of its charge, being repelled by the negative which is not on AB, passes over into the Leyden jar L. When l reaches the position o, it again becomes charged by induction, this time positively, and more strongly than at first, since now the negative on AB, as well as the positive on CD, is acting inductively upon the rod rs. When l reaches the position u, a part of its now strong positive charge passes to CD, thus increasing the positive charge upon this inductor. In the position v, the remainder of the positive charge on l passes over to L'. This completes the cycle for l. Thus, as the rotation continues, AB and CD acquire stronger and stronger charges, the inductive action upon rs becomes more and more intense, and positive and negative charges are continuously imparted to L and L' until a discharge takes place between the knobs R and S.

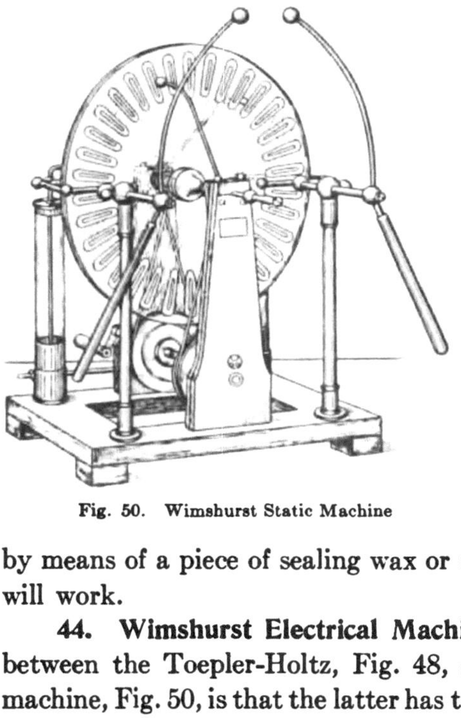

Fig. 50. Wimshurst Static Machine

There is usually sufficient charge on one of the inductors to start the machine, but in damp weather it is often found necessary to apply a charge to one of them by means of a piece of sealing wax or a glass rod before the machine will work.

44. Wimshurst Electrical Machine. The essential difference between the Toepler-Holtz, Fig. 48, and the Wimshurst electrical machine, Fig. 50, is that the latter has two plates revolving in opposite directions and that these plates carry a large number of tinfoil strips which act alternately as inductors and as carriers, thus dispensing with the necessity of separate inductors. The action of the machine may be understood readily from Fig. 51. Suppose that a small negative charge is placed on a. This, acting inductively on the rod rs, charges a' positively. When a' in the course of the rotation reaches the position b', it acts inductively upon the rod

$s'r'$ and thus charges the disk b negatively. It will be seen that henceforth all the disks in the inner circle receive + charges as they pass the brush r, and that all the disks in the outer circle, i.e., on the back plate, receive − charges as they pass the brush s'. Similarly, on the lower half of the plates all the disks on the inner circle receive − charges as they pass the brush s, and all the disks on the outer circle receive + charges as they pass the brush r'.

When the positive charges on the inner disks come opposite the combs c, they pass off to the + knob of the machine or to the Leyden jar connected with it. The same process is occurring on the other side, where − charges are being taken off. When a spark passes, the Leyden jars and the connecting system of conductors are restored to their initial conditions and the process begins again.

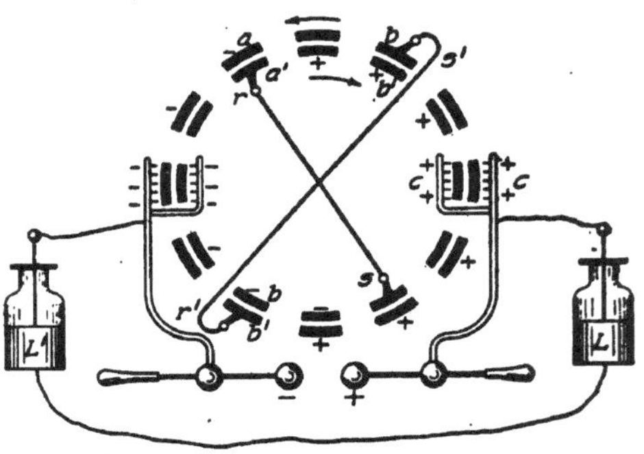

Fig. 51. Diagram of Action of Wimshurst Machine

EXAMPLES FOR PRACTICE

1. If an electrified rod is brought near to a pith ball suspended by a silk thread, the ball is first attracted to the rod and then repelled from it. Explain.

2. Two like metal spheres are charged with +20 and −10 electrostatic units of electricity. With what force will they attract each other when placed 5 cm. apart in air? Ans. 8 dynes

3. If the two spheres of the previous problem are made to touch and are then returned to their former positions, with what force will they act on each other? Will this force be attraction or repulsion? Ans. 1 dyne. Repulsion

4. If you hold a brass rod in the hand and rub it with silk, the rod will show no sign of electrification; but if you hold the brass rod with a piece of sheet rubber and then rub it with silk, you will find it electrified. Explain.

5. If you are given a positively charged insulated sphere, how could you charge two other spheres, one positively and the other negatively, without diminishing the charge on the first sphere?

6. Given a gold-leaf electroscope, a glass rod, and a piece of silk, how, in general, would you proceed to test the sign of the electrification of an unknown charge?

7. When a negatively electrified cloud passes over a house provided with a lightning rod, positive electricity passes into the cloud and negative electricity to the rod. Where do these charges which pass into the cloud and to the rod come from, and what effect do they have on the p.d. between the cloud and the earth?

8. If an electroscope is placed in a cage of wire netting and an electrified glass rod brought near it, will any effect be observed? Explain.

9. The heating coil of a certain electric teakettle has a p.d. of 110 volts applied to it and consumes energy at the rate of 440 watts. How many electrostatic units of electricity pass through the coil per second? To how many *elementary* charges (atoms of electricity) is this equivalent?

Ans. $\begin{cases} 12 \times 10^9 \text{ electrostatic units} \\ 24 \times 10^{18} \text{ atoms of electricity} \end{cases}$

10. Why is the capacity of a conductor greater when another conductor connected to the earth is near it, than when it stands alone?

ELECTRICITY IN MOTION—ELECTRICAL CURRENTS

45. Magnetic Effect Due to a Charge in Motion. An electrical charge at rest produces no magnetic effort whatever. This can be proved by bringing a charged body near a compass needle or a suspended magnet. It will attract both ends equally well by virtue of the principle of electrostatic induction. If the effect were magnetic, one end should be repelled and the other attracted. Again, if a sheet of zinc, aluminum, or copper is inserted between the deflected needle and the charge, all effect which was produced upon the needle by the charge will be cut off, for the metallic sheet will act as an electric screen, cf. Section 35. But if such a metal screen is inserted between a compass needle and a magnet, its insertion has no effect at all on the magnetic forces, cf. Section 6.

If, however, a charged Leyden jar is discharged through a coil which surrounds an unmagnetized knitting needle in the manner shown in Fig. 52, the needle will be found after the discharge to have become distinctly magnetized. If the sign of the charge on the jar is reversed, the poles will, in general, be reversed also.

This experiment demonstrates the existence of some connection between electricity and magnetism. Just what this connection is, is not yet known with certainty; but it is known that *magnetic effects are always observable near the path of a moving electrical charge,* while no such effects can ever be observed near a charge at rest.

An electrical charge in motion is called an electrical current, and the presence of such a current in a conductor is most commonly detected by the magnetic effect which it produces.

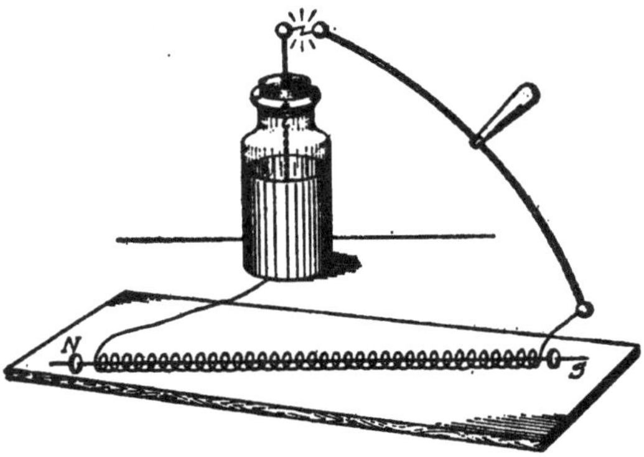

Fig. 52. Experiment to Prove That a Moving Static Charge Is a Current

46. Galvanic Cell. When a Leyden jar is discharged, but a very small quantity of electricity passes through the connecting wires, since the current lasts but a small fraction of a second. If we could keep the current flowing continuously through the wire, we should expect the magnetic effect to be more pronounced. This might be done by discharging Leyden jars in rapid succession through the wire. In 1786, however, Galvani, an Italian anatomist at the University of Bologna, accidentally discovered that there is a chemical method for producing such a continuous current. His discovery was not understood, however, until Volta, professor of physics at Como, devised an arrangement which is now known sometimes as the *voltaic,* sometimes as the *galvanic* cell.

Such a cell consists in its simplest form of a strip of copper and a strip of zinc immersed in dilute sulphuric acid, Fig. 53. If the wires leading from the copper and the zinc are connected for a few seconds to the end of the coil of Fig. 52, when an unmagnetized needle lies within this coil, the needle will be found to be much more strongly magnetized than it was when the Leyden jar was discharged through the coil. Or, if the wire connecting the copper and zinc is simply held above the needle in the manner shown in Fig. 54, the latter will be found to be strongly deflected. It is evident from these experiments that the wire which connects the terminals of a galvanic cell carries a current of electricity. Historically, the second of these experiments, performed by the Danish physicist Oersted in 1819, preceded the discovery of the magnetizing effect of currents upon needles. It created a great deal of excitement at the time because it was the first clew to the relationship between electricity and magnetism which had been found.

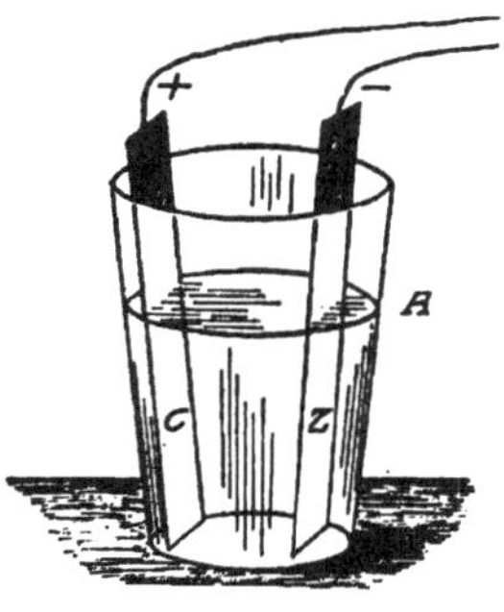

Fig. 53. Simple Galvanic Cell

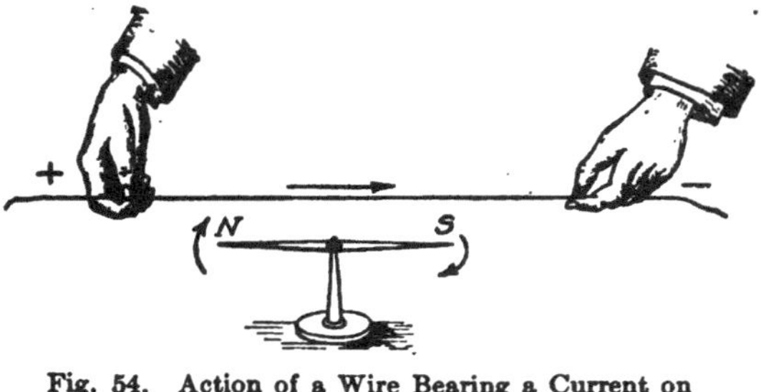

Fig. 54. Action of a Wire Bearing a Current on Magnetic Needle

It might be inferred from the above experiments that the two plates of a galvanic cell when not connected by a wire, carry static positive and negative charges just as do the two coats of a Leyden jar before it is discharged through the wire. This inference can be easily verified with an electroscope.

Thus, if a metal plate *A*, Fig. 55, covered with shellac on its lower side and provided with an insulating handle, is placed upon a similar plate *B* which is in contact with the knob on an electroscope; and if the copper plate, for example, of a galvanic cell is connected to *A* and the zinc to *B*; then, when the connecting wires are removed and the plate *A* lifted away from *B*, the leaves of the electroscope will diverge and when tested will be found to be negatively charged. If the deflection observed in the leaves of the electroscope is too small for purposes of demonstration, the conditions can be bettered by using a battery of from five to ten cells instead of the single cell.

If, however, the plates *A* and *B* are sufficiently large—say three or four inches in diameter—and if their surfaces are very flat, a single cell will be found to be sufficient. If, on the other hand, the copper plate is connected to *B* and the zinc to *A* in the above experiment, the electroscope will be found to be positively charged. This shows clearly that the copper plate possesses a positive electrical charge, while the zinc plate possesses a negative charge, these charges originating in the chemical action within the galvanic cell.

In this experiment, the two metal plates separated by shellac constitute an electrical condenser which is charged positively on the one side and negatively on the other by connecting it with the two plates of the galvanic cell, in precisely the same way in which a Leyden jar is charged by connecting its two coats, one to one terminal and the other to the other terminal of a static machine. The potential of plate *B* is increased by moving *A* away from it, just as in the arrangement shown in Fig. 44, the potential of *A* was increased by moving *B* away from it. This device makes it possible to detect very small potential differences.

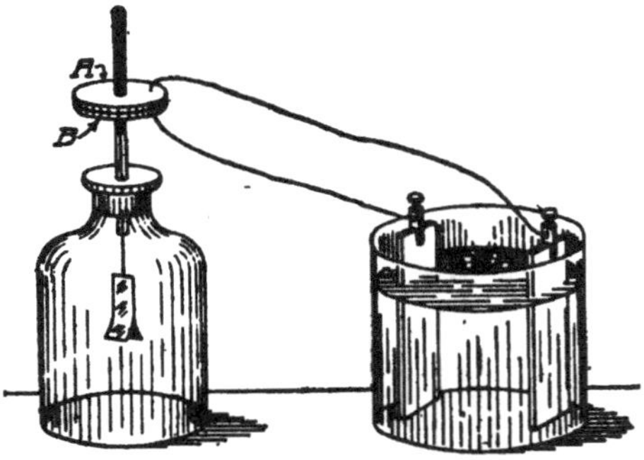

Fig. 55. Experiment Showing Method of Charging Electroscope by Battery Current

47. Comparison of a Galvanic Cell with a Static Machine. If one of the terminals of a galvanic cell is touched directly to the knob of the gold-leaf electroscope without the use of the condenser plates *A* and *B* of Fig. 55, no divergence of the leaves can be detected, but if one knob of a static machine in operation were so touched, the leaves would be thrown apart very violently. Since we have seen that the divergence of the leaves is a measure of the potential of the body to which they are connected, we learn from this experiment that the chemical actions going on in a galvanic cell are able to produce between its terminals but very small potential differences in comparison with that produced by the static machine between its terminals. As a matter of fact, the potential difference between the terminals of the cell is but one volt, cf. Section 53, while that between the terminals of an electrical machine may be several hundred thousand volts.

On the other hand, if the knobs of the static machine are connected to the ends of the wire shown in Fig. 54, and the machine operated, the current will not be large enough to produce any appreciable effect upon the needle. Since, under these same circumstances the galvanic cell produced a very large effect upon the needle, we learn that although the cell develops a much smaller p.d. than does the static machine, it nevertheless sends through the wire a very much larger amount of electricity per second. This means merely that the chemical actions which are going on within the cell are able to recharge the plates, when they become discharged through the electric wire, far more rapidly than the static machine is able to recharge its terminals after they have once been discharged.

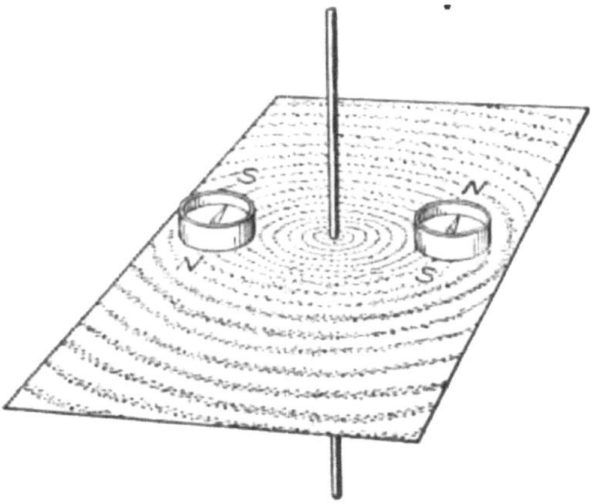

Fig. 56. Magnetic Field about a Wire Bearing a Current

48. Shape of Magnetic Field about a Current. If we place the wire which connects the plates of a galvanic cell in a vertical position, Fig. 56, and explore with a compass needle the shape of the magnetic field about the current, we find that the magnetic lines are concentric circles lying in a plane perpendicular to the wire and having the wire as their common center. If we reverse the direction of the current, we find that the direction in which the compass needle points reverses also. If the current is very strong, say 40 amperes (Section 50), this shape of the field can be shown by scattering iron filings on a plate through which the current passes, in the manner shown in Fig. 56. The relation between the direction in which the current flows and the direction in which the positive end of the needle points (this is, by definition, the direction of the magnetic field) is given in the following convenient rule: *If the right hand grasps the wire as in Fig. 57, so that the thumb* *points in the direction in which the positive electricity is moving, that is, in the direction from the copper toward the zinc, then the magnetic lines encircle the wire in the same direction as do the fingers of the hand.* Another way of stating this rule is as follows: *The relation between the direction of the current in a wire and the direction of the magnetic lines about it, is the same as the relation between the direction of the forward motion of a right-hand screw and the direction of rotation when it is being driven in.* In this form the rule is known as the *right-hand screw rule.*

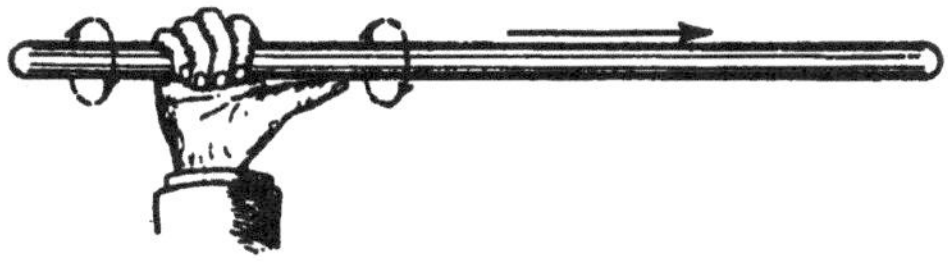

Fig. 57. Direction of Flow of Magnetic Lines about a Wire

Fig. 58 shows the shape of the magnetic field about a wire in which the current is flowing toward the observer, "out" from the plane of the figure, and Fig. 59 shows the shape of the magnetic field when the current is flowing away from the observer, "in" or back from the plane of the figure.

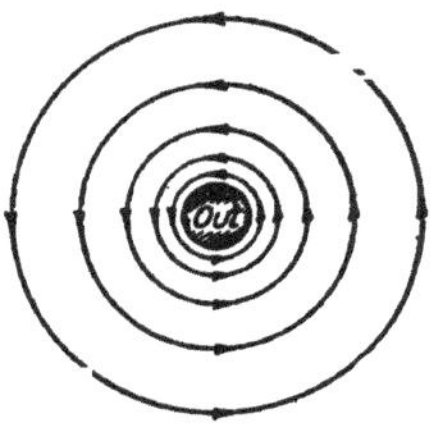

Fig. 58. Field about a Wire. Current Flowing toward Observer

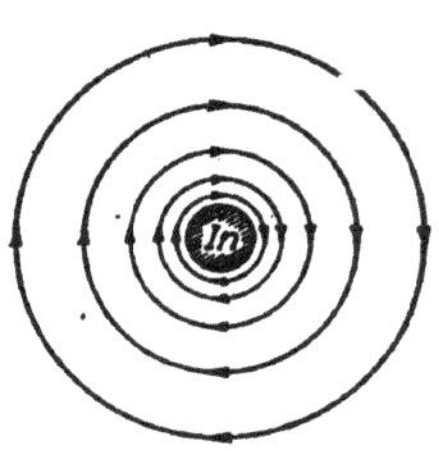

Fig. 59. Field about a Wire. Current Flowing away from Observer

49. Galvanometers. When the terminals of a simple cell are connected to the ends of a coil which passes around a magnetic needle as in Fig. 60, the needle will be deflected more strongly than when a single wire was used, as in the experiment illustrated in Fig. 54, for, in accordance with the right-hand rule, both the upper and the lower portions of the coil tend to make the needle turn in the same direction. It will come to rest in a position at right angles to the plane of the coil.

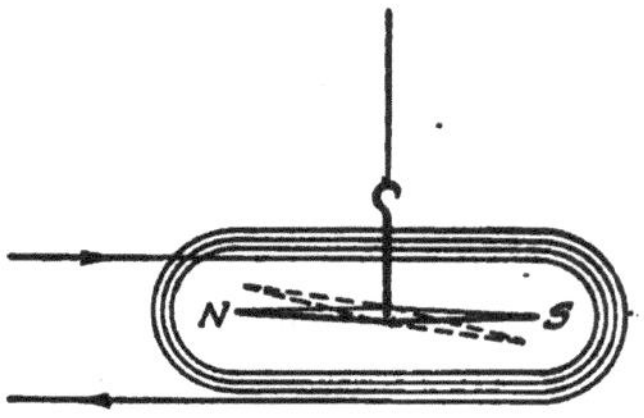

Fig. 60. Simple Galvanometer

Again, let a coil of say 200 turns of No. 30 copper wire be suspended between the poles of a magnet *NS*, as in Fig. 61. The suspending wires should be cf No 40 copper wire. As soon as the current from the galvanic cell is sent through the coil, it will turn so as to set itself at right angles to the line connecting *N* and

S, i.e., at right angles to the lines of force of the magnet. The only essential difference between the two experiments is that in the first the coil is fixed and the magnet free to turn, while in the second the magnet is fixed and the coil is free to turn. In both cases, the passage of the current tends to cause the suspended system to rotate through 90 degrees.

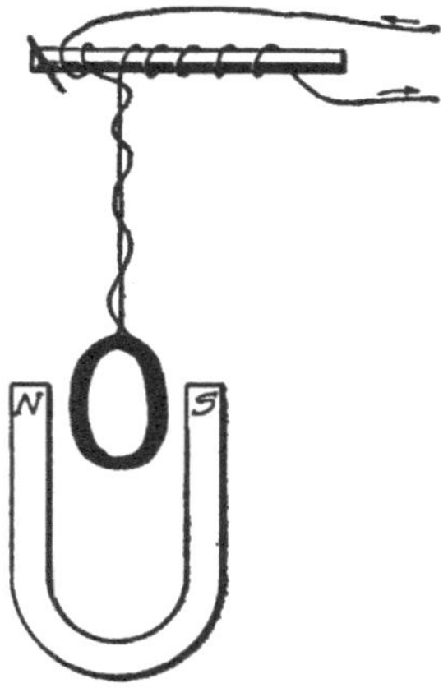

Fig. 61. Principle of D'Arsonval Galvanometer

These two experiments illustrate the principle underlying nearly all current-measuring instruments or galvanometers. Instruments of the suspended-coil type are the more common and the more convenient. They are usually called D'Arsonval galvanometers. Fig. 62 represents one form of such a galvanometer used for laboratory experiments, and Fig. 63 shows a somewhat modified form suitable for classroom experiments. The current in passing through such an instrument enters through the suspending wire, encircles the coil, and then

Fig. 62. Commercial D'Arsonval Galvanometer
Courtesy of Wm. Gaertner & Company, Chicago

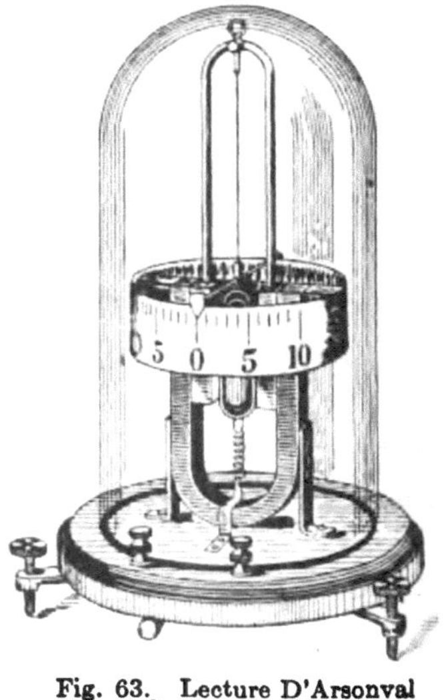

Fig. 63. Lecture D'Arsonval Galvanometer

passes out through the fine spiral coil shown below the movable coil, or vice versa. The strength of the field in which the coil rotates is considerably increased by placing a soft iron cylinder in the space inside the movable coil, thus causing the instrument to respond to very much smaller currents.

MEASURING OF ELECTRIC CURRENT

50. Unit of Current—the Ampere. Ampère was the first investigator who made quantitative measurements on continuous currents by means of their magnetic effects. Hence the practical unit of current is named, in honor of him, the *ampere*. It may be defined as the current which, when flowing through a circular coil of 100 turns and a radius of 10 centimeters, will produce at its center a magnetic field of strength equal to 2π dynes. Fig. 64 shows the shape of the magnetic field about such a coil. The ampere is approximately the same as the current which, flowing through a circular coil of 3 turns and a radius of 10 centimeters, set in a north-and-south plane, will produce a deflection of 45 degrees in a small compass needle placed at its center as in Fig. 64.

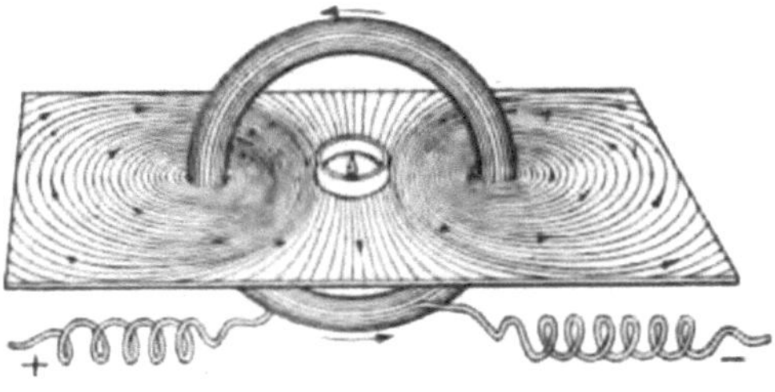

Fig. 64. Magnetic Field about a Coil

A very ingenious method used by Lord Rayleigh for measuring the strength of an electric current is shown in Fig. 65. The coils *A*, *C*, and *D* are connected in series, that is, so that the same current which is being measured passes through each coil successively. Coil *A* is first balanced by weights placed in pan *E*. The current to be measured is then sent

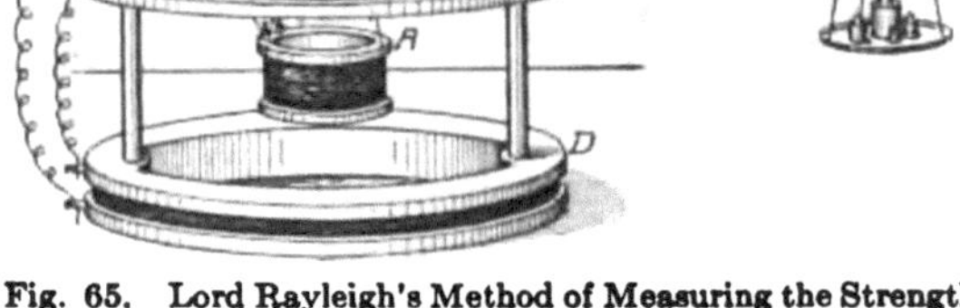

Fig. 65. Lord Rayleigh's Method of Measuring the Strength of an Electric Current

through the coils in such a way that coil A is attracted by coil D and repelled by coil C and the total force due to the magnetic action between the coils is then equal to the weight which must be added to pan E in order to maintain a balance. This force can also be calculated in terms of the current and the dimensions of the coil. Hence from the measured force, i.e., the weight added to pan E to maintain a balance, and the dimensions of the coil, the current can be calculated.

By sending the same current which passed through the coils of Fig. 65, also through a plating solution of silver nitrate, Lord Rayleigh found that a current of *one ampere deposited* 0.0011179 *gram of silver in one second. By international agreement, the ampere is now defined as the current which will deposit silver at the rate of* 0.001118 *gram per second.*

51. Commercial Ammeter. Fig. 66 shows the construction of the usual form of ammeter, which is really nothing but a ***shunted moving-coil galvanometer.*** The coil *c* is pivoted on jewel bearings and is held at its zero position by a spiral spring *p*. When a current flows through the instrument, were it not for the spring *p* the coil would turn through about 120 degrees, or until its N pole came opposite the S pole of the magnet. This zero position is chosen because with pole pieces shaped as in the figure, it enables the scale divisions to be nearly equal. The coil *c* is always of high resistance compared to the coils *r*, and the coils *r* have a very small resistance. By connecting the coils *r* (shunt coils) in the proper way, the instrument is sometimes made so that the same full-scale deflection may read, for example, 1.5, 15.0, or 30 amperes.

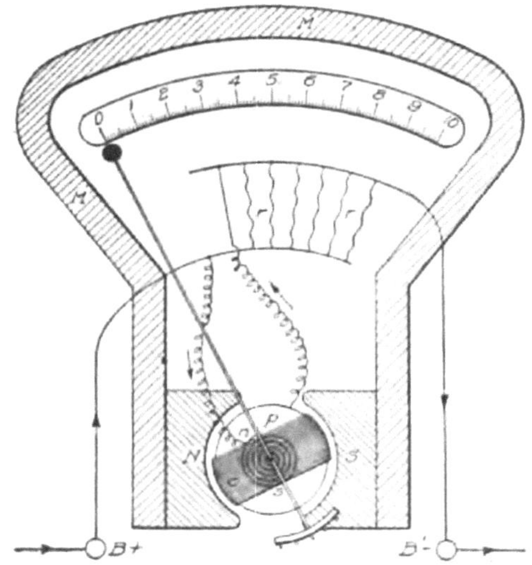

Fig. 66. Diagram of Ammeter

52. Illustrations of the Ampere. When the electrodes of a new dry cell are touched to an ammeter, a current of from 20 to 30 amperes will flow. A telegraph relay and a telephone will operate on about one-tenth of an ampere. Electrical welders require from a few amperes to several hundred, depending on the character of the work. A 16-candle-power tungsten lamp takes about .18 of an ampere to light it while a 16-candle-power carbon filament lamp takes about .5 of an ampere. Fan motors, sewing-machine motors, and washing-machine motors operated on 110 volts take less than 1 ampere, while street-car motors require about 50 amperes.

EXAMPLES FOR PRACTICE

1. How could you test whether or not the strength of an electric current is the same in all parts of a circuit?

2. Under what conditions will an electric charge produce a magnetic effect?

3. How would you prove that a static charge does not produce a magnetic effect?

4. In what direction will the north pole of a magnetic needle be deflected if it is held above a current flowing from north to south?

5. A man stands beneath a north-and-south trolley line and finds that a magnetic needle in his hand has its north pole deflected toward the east. What is the direction of the current flowing in the wire?

6. A loop of wire lying on the table carries a current which flows around it in a clockwise direction. Would a north magnetic pole at the center of the loop tend to move up or down?

7. How would you make a simple galvanometer that would enable you to tell which of a number of cells would give the most current?

53. Electromotive Force and Its Measurements. The potential difference which a galvanic cell or any other generator of electricity is able to maintain between its terminals when these terminals are not connected by a wire, i.e., the total electrical pressure which the generator is capable of exerting, is commonly called its electromotive force, usually abbreviated e.m.f. *The e.m.f. of an electrical generator may then be defined as its capacity for producing electrical pressure, or p.d.* This p.d. might be measured, as in Section 38 by the deflection produced in an electroscope, or an electrostatic voltmeter, when one terminal was connected to the case of either instrument and the other terminal to the knob. Potential differences are in fact measured in this way in all so-called electrostatic voltmeters, which are now coming more and more into use.

The more common type of potential-difference measurers, so-called *voltmeters*, consists, however, of an instrument made similar to an ammeter, save that the voltmeter does not have a shunt coil, but, instead, has a coil made of an enormous number of turns of extremely fine wire placed in series with the movable coil, so that it carries a very small current, (see Fig. 70).

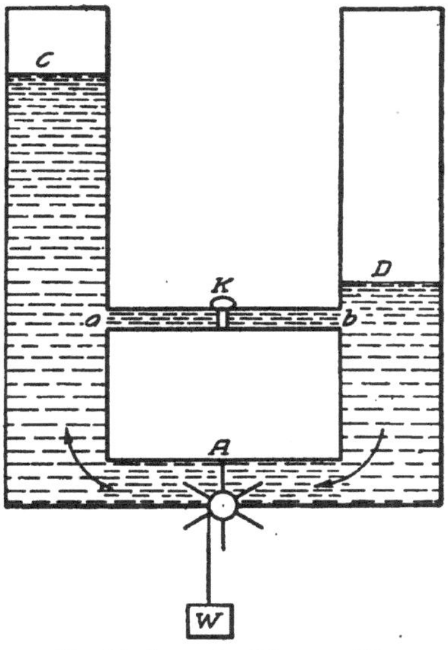

Fig. 67. Hydrostatic Diagram Illustrating Effect of Potential Difference

The amount of current which it does carry, however, and therefore the amount of deflection of its needle is taken as proportional to the difference in electrical pressure existing between its ends, when these are touched to the two points whose p.d. is sought. The principle underlying this type of voltmeter will be better understood from a consideration of the following water analogy. If the stopcock *K*, Fig. 67, in the pipe connecting the water tanks *C* and *D*, is closed, and if the water wheel *A* is set in motion by applying a weight *W*, the wheel will turn until it creates such a difference in the water levels between *C* and *D* that the back pressure against the left face of the wheel stops it and brings the weight *W* to rest. In precisely the same way, the chemical action within the galvanic cell whose terminals are not joined, Fig. 68, develops positive and negative charges upon these terminals, that is, creates a p.d. between them, until the back electrical pressure through the cell due to this p.d. is sufficient to put a stop to further chemical action.

Now, if the water reservoirs, Fig. 67, are put in communication by opening the stopcock *K*, the difference in level between *C* and *D* will begin to fall, and the wheel will begin to build it up again. But if the carrying capacity of the pipe *ab* is small in comparison with the capacity of the wheel to remove water from *D* and to supply it to *C*, then the difference of level which permanently exists between *C* and *D* when *K* is open will not be appreciably smaller than when it is closed. In this case the current which flows through *AB* may obviously be taken as a measure of the difference in pressure which the pump is able to maintain between *C* and *D* when *K* is closed.

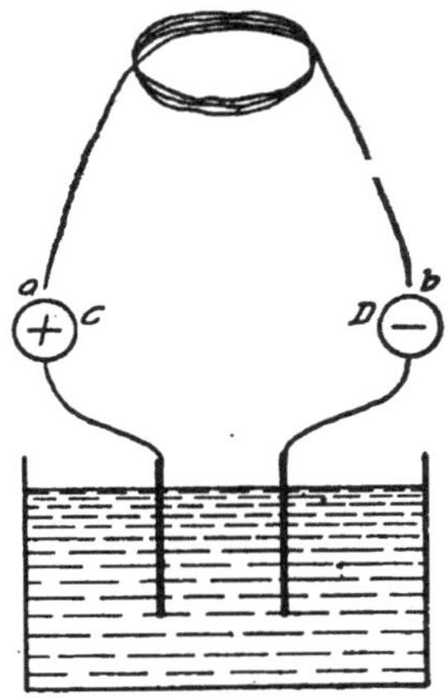

Fig. 68. Chemical Action in Simple Cell Creates a P. D. in Circuit

In precisely the same way, if the terminals *C* and *D* of the cell, Fig. 68, are connected by attaching to them the terminals *a* and *b* of any conductor, they at once begin to discharge through this conductor, and their p.d. therefore begins to fall. But if the chemical action in the cell is able to recharge *C* and *D* very rapidly in comparison with the ability of the wire to discharge them, then the p.d. between *C* and *D* will not be appreciably lowered by the presence of the connecting conductor. In this case the current which flows through the conducting coil, and therefore the deflection of the needle at its center, may be taken as a measure of the electrical pressure developed by the cell, that is, of the p.d. between its unconnected terminals.

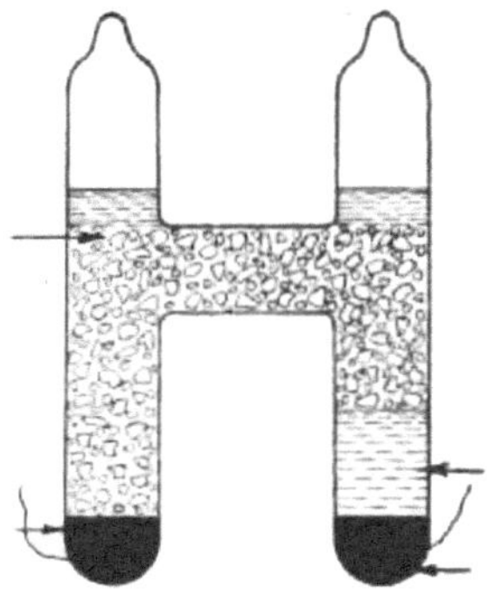
Fig. 69. Construction of Weston Standard Cell

54. Volt. On account of the fact that the volt as defined in Section 37 is so difficult to reproduce or to use as a standard, it is highly desirable to have a *standard p.d.* which can be easily and accurately reproduced. For this purpose the Weston normal cell has been chosen. This cell, Fig. 69, consists of a positive electrode of mercury in a paste of mercurous sulphate, and a negative electrode of cadmium amalgam in a saturate solution of cadmium sulphate. It is so easily and exactly reproducible and has an e.m.f. of such extraordinary constancy, that it has been taken by international agreement as the standard in terms of which all e.m.f.'s and p.d.'s are rated. Careful experiments show that the e.m.f. of a Weston normal cell at 20° C., measured in volts as defined in Section 37, is 1.0183 volts.

The legal definition of the volt is, then, an electrical pressure equal to $\frac{1}{1.0183}$ *of that produced by a Weston normal cell at 20° C.*

55. Commercial Voltmeter. Fig. 70 shows the construction of the usual type of voltmeter, which is really a *high-resistance moving-coil galvanometer.* It differs from the ammeter of Section 51 in the following respects: Instead of the shunt coils *r* which are in parallel with the moving coil *c*, a coil of high resistance *R* is placed in series with *c* so that the instrument, instead of having a negligible resistance as in the case of the ammeter, has a high resistance. The full-scale deflection can be made to represent different voltages such as 1.0, 10, or 150 volts by using coils similar to *R* placed in series with the instrument.

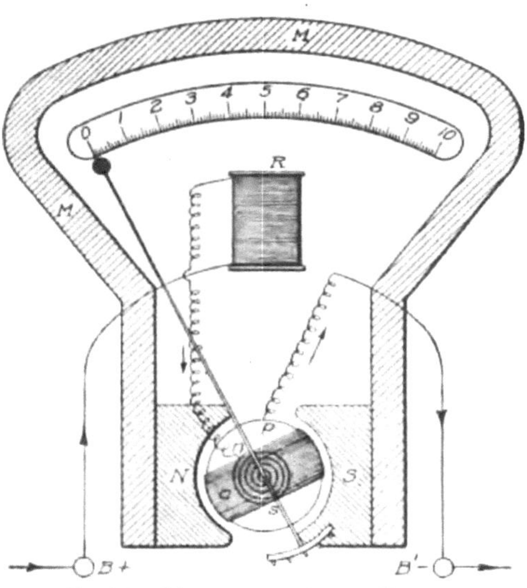

Fig. 70. Diagram of Voltmeter Principle

56. Illustrations of the Volt. The Weston Standard cell, Fig. 69, produces an e.m.f. of 1.0183 volts at 20° C., a dry cell about 1.45 volts, and a lead storage cell about 2.1 volts. The p.d. between two light mains is usually 110 volts but is often 220 volts. The p.d. between a street-car trolley line and the earth is about 550 volts. The p.d. between the electrodes of a spark-coil used to transmit transatlantic radiograms is several million volts.

57. Electromotive Forces of Galvanic Cells. When a voltmeter of any sort is connected to the terminals of a galvanic cell, it is found that the deflection produced is altogether independent of the shape or size of the plates or of their distance apart. But if the nature of the plates is changed, the deflection changes. Thus, while copper and zinc in dilute sulphuric acid have an e.m.f. of one volt, carbon and zinc show an e.m.f. of at least 1.5 volts, while carbon and copper will show an e.m.f. of very much less than a volt. Similarly, by changing the nature of the liquid in which the plates are immersed, we can produce changes in the deflection of the voltmeter. We learn therefore that *the e.m.f. of a galvanic cell depends simply upon the materials of which the cell is composed and not at all upon the shape, size, or distance apart of the plates.*

58. Electrical Resistance. If the terminals of a galvanic cell are connected first to ten feet of No. 30 copper wire, and then to ten feet of No. 30 German-silver wire, it is found that a compass needle placed at a given distance from the copper wire will show a much larger deflection than when placed the same distance from the German-silver wire. A cell, therefore, which is capable of developing a certain fixed electrical pressure is able to force very much more current through a given wire of copper than through an exactly similar wire of German silver. We say, therefore, that German silver offers a higher *resistance* to the passage of electricity than does copper. Similarly, every particular substance has its own characteristic power of transmitting electrical currents. Silver is the best conductor of any of the known substances. The resistances of different substances are commonly referred to silver as a standard, and the ratio between the resistance of a given wire of any substance and the resistance of an exactly similar silver wire is called the *specific resistance* of that substance. The specific resistances of some of the commoner metals are given below:

Silver	1.00	Soft iron	7.40	German silver, 18 per cent nickel	18.0
Copper	1.13	Nickel	7.87	Hard steel	21.0
Aluminum	2.00	Platinum	9.00	Mercury	62.7

The unit of resistance is the resistance at 0 degrees C of a column of mercury 106.3 cm. long and 1 sq. mm. in cross-section. It is called an ohm, in honor of the great German physicist, Georg Ohm (1789-1854). A length of 9.35 feet of No. 30 copper wire, or 6.2 inches of No. 30 German-silver wire, has a resistance of about one ohm. Copper wire of the size shown in Fig. 71 has a resistance of about 2.62 ohms per mile.

Fig. 71. Copper Conductor

The resistances of all metals increase with a rise in temperature. The resistances of liquid conductors, on the other hand, usually decrease with a rise in temperature. Carbon and a few other solids show a similar behavior: the carbon-filament incandescent lamp has only about half the resistance when hot that it has when cold. The resistances of wires of the same material are found to be directly proportional to their lengths, and inversely proportional to their cross-sections.

59. Ohm's Law. In 1827 Ohm announced the discovery that *the currents furnished by different galvanic cells, or combinations of cells, are always directly proportional to the e.m.f.'s existing in the circuits in which the currents flow, and inversely proportional to the total resistances of these circuits;* i.e., if I represents the current in amperes, E the e.m.f. in volts, and R the resistance of the circuit in ohms, then Ohm's law as applied to the complete circuit is:

$$I = \frac{E}{R} \tag{1}$$

that is, current equals electromotive force divided by resistance.

As applied to any portion of an electrical circuit, Ohm's law is

$$I = \frac{\text{p.d.}}{r} \tag{2}$$

that is, current equals potential difference divided by resistance, where p.d. represents the difference of potential in volts between any two points in the circuit, and r the resistance in ohms of the conductor connecting these two points. This is one of the most important laws in physics.

Both of the above statements of Ohm's law are included in the equation:

$$\text{amperes} = \frac{\text{volts}}{\text{ohms}} \tag{3}$$

60. Internal Resistance of a Galvanic Cell. If the zinc and the copper plates of a simple galvanic cell are connected to an ammeter, and the distance between the plates then increased, the deflection of the needle is found to decrease; or if the amount of immersion is decreased, the current also will decrease. But since the e.m.f. of a cell was shown in Section 57 to be wholly independent of the area of the plates immersed or of the distance between them, it will be seen from Ohm's law that the change in the current in these cases must be due to some change in the total resistance of the circuit.

Since the wire which constitutes the outside portion of the circuit has remained the same, we must conclude that *the liquid within the cell, as well as the external wire, offers resistance to the passage of the current.* This internal resistance of the liquid is directly proportional to the distance between the plates, and inversely proportional to the area of the immersed portion of the plates. If, then, we represent the external resistance of the circuit of a galvanic cell by R_e and the internal by R_i, then Ohm's law as applied to the entire circuit takes the form:

$$I=\frac{E}{R_e+R_i} \quad (4)$$

Thus, if a simple cell has an internal resistance of 2 ohms and an e.m.f. of 1 volt, the current which will flow through the circuits when its terminals are connected by 9.3 ft. of No. 30 copper wire (1 ohm), is $\frac{1}{1+2}=.33$ ampere. This is about the current which is usually obtained from an ordinary Daniell cell, see Section 65.

EXAMPLES FOR PRACTICE

1. If the potential difference between the terminals of a cell on open circuit is to be measured by means of a galvanometer, why must the galvanometer have a high resistance?

2. How would you make a simple voltmeter if you had some wire and a horseshoe magnet?

3. How long a piece of No. 30 copper wire will have the same resistance as a meter of No. 30 German-silver wire? See Section 58.

4. How many ohms' resistance in each of the wires in problem 3?

5. The resistance of a certain copper wire is 1 ohm. What will be the resistance of another copper wire of the same length, but of one-half the diameter?

6. A voltmeter reads 2 volts when connected to two points 9.35 feet apart on a No. 30 copper wire which is carrying a current. What is the strength of the current in amperes?

7. In Example 6 what would the voltmeter read if attached to two points which were 1.87 feet apart?

8. A current which is passed through an ammeter and then through a silver-plating solution for 25 minutes is found to deposit

1.677 grams of silver. During this time the ammeter read .95 ampere. Does the ammeter read too high or too low at this part of the scale and how much? (This is a standard method for testing an ammeter.)

9. The current from a Daniell cell, Section 65, is sent through a 2-ohm coil and an ammeter of negligible resistance. If the ammeter reads 0.27 ampere and if the e.m.f. of the cell is 1.08 volts, find the internal resistance of the cell in ohms.

PRIMARY CELLS

61. Action of a Simple Cell. If the simple cell already mentioned—namely, zinc and copper strips in dilute sulphuric acid—is carefully observed, it will be seen that, so long as the plates are not connected by a conductor, fine bubbles of gas are slowly formed at the zinc plate, but none at the copper plate. As soon, however, as the two strips are put into electrical connection, bubbles instantly appear in great numbers about the copper plate and at the same time a current manifests itself in the connecting wire, Fig. 72. The bubbles are of hydrogen. Their original appearance on the zinc plate may be prevented either by using a plate of chemically pure zinc, or by amalgamating impure zinc, that is, by coating it over with a thin film of mercury. But the bubbles on the copper cannot be thus disposed of. They are an invariable accompaniment of the current in the circuit. If the current is allowed to run for a considerable time, it will be found that the zinc wastes away, even though it has been amalgamated, but the copper plate does not undergo any change.

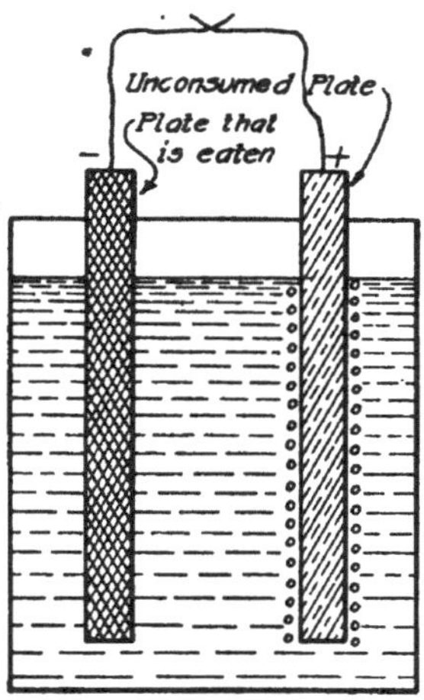

Fig. 72. Diagram Illustrating Action of Simple Cell

An electrical current in a simple cell is, then, accompanied by the eating up of the zinc plate by the liquid, and by the evolution of hydrogen bubbles at the copper plate. In every type of galvanic cell, actions similar to these two are always found. That is, *one of the plates is always eaten up, and on the other some element is deposited.* The plate which is eaten is always the one which is found to be negatively charged, while the other is always found to be positively charged; so that in all galvanic cells, when the terminals are connected through a wire, the positive electricity flows through this wire from the uneaten plate to the eaten plate. It will be remembered that the direction in which the *positive* electricity flows is taken for convenience as the direction of the current, see Sections 25 and 36.

62. Theory of Action of a Simple Cell. A simple cell may be made of any two dissimilar metals immersed in a solution of any acid or salt. For simplicity, let us examine the action of a cell composed of plates of zinc and copper immersed in a dilute solution of hydrochloric acid. The chemical formula for hydrochloric acid is HCl. This means that each molecule of the acid consists of one atom of hydrogen combined with one atom of chlorine. In accordance with the theory now in vogue among physicists and chemists, when hydrochloric acid is mixed with water so as to form a dilute solution, the HCl molecules split up into two electrically charged parts, called ions, the hydrogen ion carrying a positive charge and the chlorine ion an equal negative charge, Fig. 73. This phenomenon is known as *dissociation.* The solution as a whole is neutral; i.e., it is uncharged, because it contains just as many positive as negative ions.

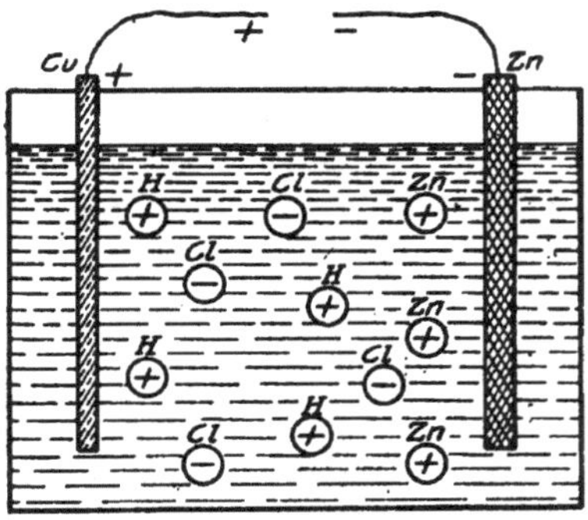

Fig. 73. Passage of Ions through the Electrolite of a Simple Cell

When a zinc plate is placed in such a solution, the acid attacks it and pulls zinc atoms into solution. Now, whenever a metal dissolves in an acid, its atoms, for some unknown reason, go into solution bearing little positive charges. *The corresponding negative charges must be left on the zinc plate* in precisely the same way in which a negative charge is left on silk when positive electrification is produced on a glass rod by rubbing it with the silk. It is in this way, then, that we attempt to account for the negative charge which we find upon the zinc plate in the experiment described in Section 46.

The passage of positively charged zinc ions into solution gives a positive charge to the solution about the zinc plate, so that the hydrogen ions tend to be repelled toward the copper plate. When these repelled hydrogen ions reach the copper plate, some of them give up their charges to it and then collect as bubbles of hydrogen gas. It is in this way that we account for the positive charge which we find on the copper plate in the experiment described in Section 46.

If the zinc and copper plates are not connected by an outside conductor, this passage of positively charged zinc ions into solution continues but a very short time, for the zinc soon becomes so strongly charged negatively that it pulls back on the plus zinc ions with as much force as the acid is pulling them into solution. In precisely the same way, the copper plate soon ceases to take up any more positive electricity from the hydrogen ions, since it soon acquires a large enough plus charge to repel them from itself with a force equal to that with which they are being driven out of solution by the positively charged zinc ions. It is in this way that we account for the fact that on open circuit no chemical action goes on in the simple galvanic cell, the zinc and copper plates simply becoming charged to a definite difference of potential which is called the e.m.f. of the cell.

When, however, the copper and zinc plates are connected by a wire, a current at once flows from the copper to the zinc, and the plates thus begin to lose their charges. This allows the acid to pull more zinc into solution at the zinc plate, and allows more hydrogen to go out of solution at the copper plate. These processes, therefore, go on continuously so long as the plates are connected. Hence, a continuous current flows through the connecting wire until the zinc is all eaten up or the hydrogen ions have all been driven out of the solution, i.e., until either the plate or the acid has become exhausted.

63. Polarization. If the simple cell which has been described is connected to an ammeter through a resistance of about .5 ohm, i.e., by about 5 feet of No. 30 copper wire, and the deflection observed for a few minutes, it is found to produce a current of continually decreasing strength; but if the hydrogen is removed from the copper plate by taking out the plate and drying it, the deflection returns to its first value. This phenomenon is called *polarization.*

The presence of the hydrogen on the positive plate causes a diminution in the strength of the current for two reasons: *First,* since hydrogen is a nonconductor, by collecting on the plate it diminishes the effective area of the plate and therefore increases the internal resistance of the cell; *second,* by collecting upon the copper plate it lowers the e.m.f. of the cell, because it virtually substitutes a hydrogen plate for the copper plate, and we have already seen, Section 57, that a change in any of the materials of which a cell is composed changes its e.m.f.

The several forms of galvanic cells in common use differ chiefly in various devices employed either for disposing of the hydrogen bubbles or for preventing their formation. The most common types of such cells are described in the following sections.

64. Bichromate Cell. The bichromate cell, Fig. 74, consists of a plate of zinc immersed in sulphuric acid between two plates of carbon, carbon being used instead of copper because it gives a greater e.m.f. In the sulphuric acid is dissolved some bichromate of potassium or sodium, the function of which is to unite chemically with the hydrogen as fast as it is formed at the positive plate, thus preventing its accumulation upon this plate.* Such a cell has the high e.m.f. of 2.1 volts. Its internal resistance is low, from .2 to .5 ohm, since the plates are generally large and close together. It will be seen, therefore, that when the external resistance is very small it is capable of furnishing a current of from 5 to 10 amperes. Since, however, the chromic acid formed by the union of the sulphuric acid with the bichromate attacks the zinc even when the circuit is open, it is necessary to lift the zinc from the liquid by the rod *A*, when the cell is not in use. Such cells are useful where large currents are needed for a short time. The great disadvantages are that the fluid deteriorates rapidly, and that the zinc cannot be left in the liquid.

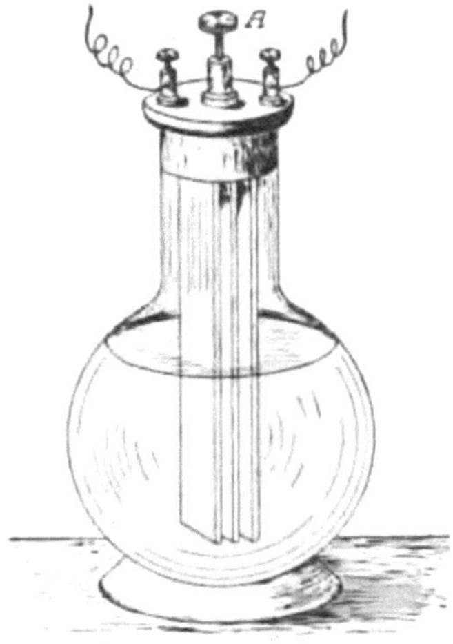

Fig. 74. Bichromate Cell

65. Daniell Cell. The Daniell cell consists of a zinc plate immersed in zinc sulphate, and a copper plate immersed in copper sulphate, the two liquids being kept apart either by means of a porous

* To set up a bichromate cell, dissolve 12 parts, by weight, of sodium bichromate in 180 parts of boiling water. After cooling, add 25 parts of commercial sulphuric acid.

earthen cup, as in the type shown in Fig. 75, or else by gravity, as in the type shown in Fig. 76. This last type, commonly called the *gravity* or *crowfoot* type, is used almost exclusively on telegraph lines. The copper sulphate, being the heavier of the two liquids, remains at the bottom about the copper plate, while the zinc sulphate remains at the top about the zinc plate.

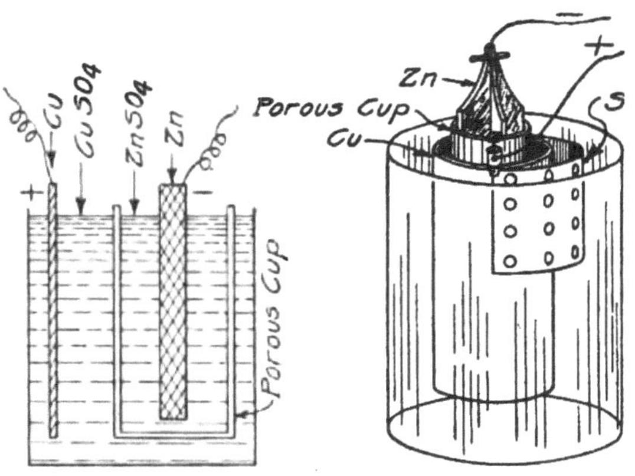

Fig. 75. Section and Complete View of Daniell Cell

In this cell, polarization is almost entirely avoided, for the reason that no opportunity is given for the formation of hydrogen bubbles. For, just as the hydrochloric acid solution described in Section 62 consists of positive hydrogen ions and negative chlorine ions in water, so the zinc sulphate ($ZnSO_4$) solution consists of positive zinc ions and negative SO_4 ions. Now the zinc of the zinc plate goes into solution in the zinc sulphate in precisely the same way that it goes into solution in the hydrochloric acid of the simple cell described in Section 62. This gives a positive charge to the solution about the zinc plate, and causes a movement of the positive ions between the two plates from the zinc toward the copper, and of negative ions in the opposite direction, both the Zn and the SO_4 ions being able to pass through the porous cup. Since the positive ions about the copper plate consist of atoms of copper, it will be seen that the material which is driven out of solution at the copper plate, instead of being hydrogen, as in the simple cell, is metallic

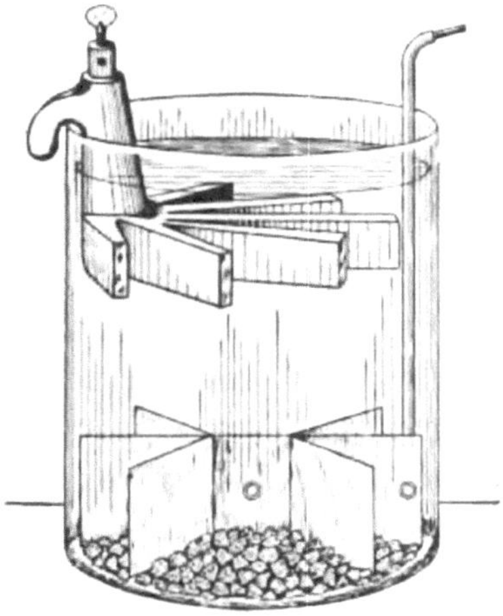
Fig. 76. Crowfoot or Gravity Cell

copper. Since, then, the element which is deposited on the copper plate is the same as that of which it already consists, it is clear that neither the e.m.f. nor the resistance of the cell can be changed because of this deposit; i.e., the cause of the polarization of the simple cell has been removed.

The advantage of the Daniell cell lies in the relatively high degree of constancy in its e.m.f. (1.08 volts). It has a comparatively high internal resistance (one to six ohms) and is therefore incapable of producing a very large current, about one ampere at most. It will furnish a very constant current, however, for a great length of time; in fact, until all of the copper is driven out of the copper sulphate solution. In order to keep a constant supply of the copper ions in the solution, copper sulphate crystals are kept in the compartment *S* of the cell of Fig. 75, or in the bottom of the gravity cell. These dissolve as fast as the solution loses its strength through the deposition of copper on the copper plate.

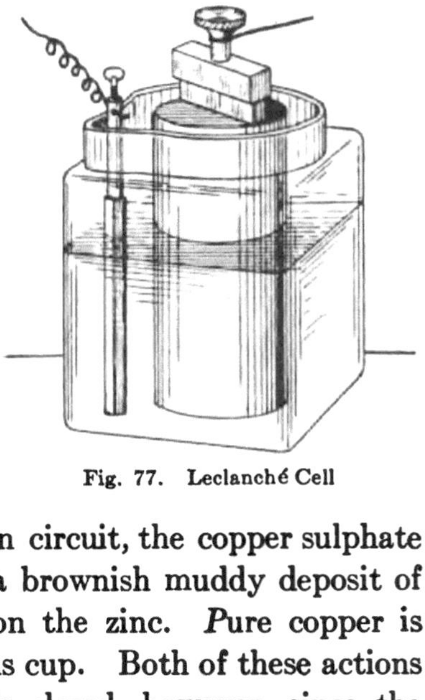

Fig. 77. Leclanché Cell

The Daniell is a so-called *closed-circuit* cell, i.e., its circuit should be left closed (through a resistance of thirty or forty ohms) whenever the cell is not in use. If it is left on open circuit, the copper sulphate diffuses through the porous cup, and a brownish muddy deposit of copper or copper oxide is formed upon the zinc. Pure copper is also deposited in the pores of the porous cup. Both of these actions damage the cell. When the circuit is closed, however, since the electrical forces always keep the copper ions moving toward the copper plate, these damaging effects are to a large extent avoided.

66. Leclanche Cell. The Leclanché cell, Fig. 77, consists of a zinc rod in a solution of ammonium chloride (150 g. to a liter of water), and a carbon plate placed inside of a porous cup which is packed full of manganese dioxide and powdered graphite or carbon. As in the simple cell, the zinc dissolves in the liquid, and hydrogen is liberated at the carbon, or positive, plate. Here it is slowly attacked by the manganese dioxide. This chemical action, however, is not quick enough to prevent rapid polarization when large currents are taken from the cell. The cell slowly recovers when allowed to stand for a while on open circuit. The e.m.f. of a Leclanché cell is about 1.5 volts, and its initial internal resistance is somewhat less than an ohm. It therefore furnishes a momentary current of from one to three amperes.

The great advantage of this type of cell lies in the fact that the zinc is not at all eaten by the ammonium chloride when the circuit is open, and that, therefore, unlike the Daniell or bichromate cells, it can be left for an indefinite time on open circuit without deterioration. Leclanché cells are used almost exclusively where momentary currents only are needed, as, for example, on doorbell circuits. The cell requires no attention for years at a time, other than the occasional addition of water to replace loss by evaporation, and the occasional addition of ammonium chloride (NH_4Cl) to keep positive NH_4 and negative Cl ions in the solution.

67. Dry Cell. The dry cell, Fig. 78, is only a modified form of the Leclanché cell. It is not really *dry*, since the carbon plate is imbedded in a moist paste contained in a cylindrical can made of zinc. The paste consists usually of crystals of ammonium chloride, three parts of plaster of Paris, one part of zinc oxide, one part of zinc chloride, and two parts of water. The plaster of Paris is used to give the paste rigidity. As in the Leclanché cell, it is the action of the ammonium chloride upon the zinc which produces the current. The dry cell must not be left on closed circuit and consequently is used largely for so-called open-circuit work, i.e., work requiring a current for short intervals of time, such as ringing doorbells, operating spark-coils for gaslighters and for exploding the gas mixture in the cylinders of gas engines on boats, aeroplanes, automobiles, etc. An old cell which has lost its strength may be greatly improved by punching a hole in the top and pouring in some water which has been saturated with ammonium chloride (sal ammoniac).

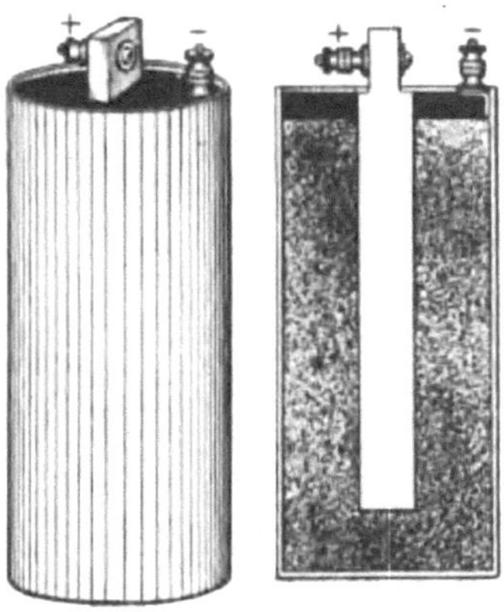

Fig. 78. Typical Dry Battery

68. Combination of Cells. There are two ways in which cells may be combined: *First,* in series; and *second,* in parallel. When they are connected in series, the zinc of one cell is joined to the copper of the second, the zinc of the second to the copper of the third, etc., the copper of the first and the zinc of the last being joined to the ends of the external resistance, Fig. 79. The e.m.f. of such a combination is the sum of the e.m.f.'s of the single cells. The internal resistance of the combination is also the sum of the internal resistances of the single cells. Hence, if the external resistances are very small, the current furnished by the combination will not be larger than that furnished by a single cell, since the total resistance of the circuit has been increased in the same ratio as the total e.m.f. But if the external resistance is large, the current produced by the combination will be very much greater than that produced by a single cell. Just how much greater can always be determined by applying Ohm's law, for if there are n cells in series, and E is the e.m.f. of each cell, the total e.m.f. of the circuit is nE. Hence if R_e is the external resistance and R_i the internal resistance of a single cell, then Ohm's law gives

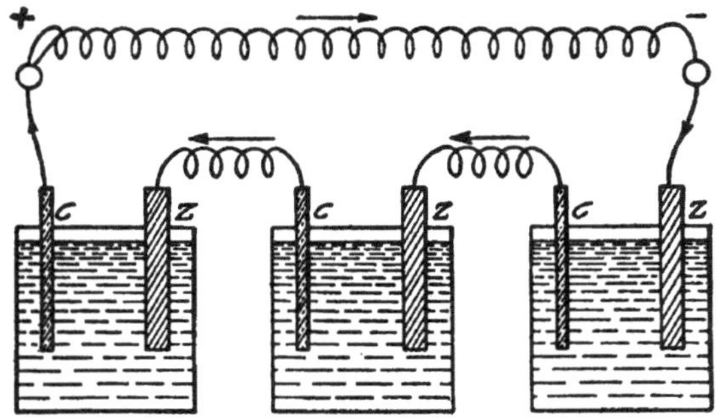

Fig. 79. Diagram of Cells in Series

$$I=\frac{nE}{R_e+nR_i}$$

If the n cells are connected in parallel, that is, if all the coppers are connected and all the zincs, as in Fig. 80, the e.m.f. of the combination is only the e.m.f. of a single cell, while the internal resistance is $\frac{1}{n}$ of that of a single cell, since connecting

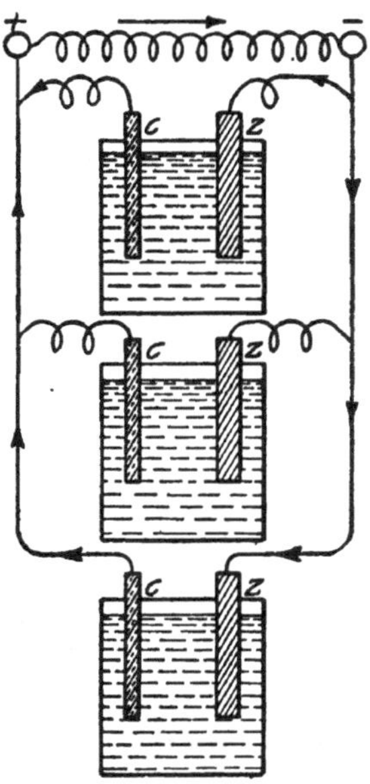

Fig. 80. Diagram of Cells in Parallel

the cells in this way is simply the equivalent to multiplying the area of the plates n times. The current furnished by such a combination will be given by the formula:

$$I=\frac{E}{R_e+\frac{R_i}{n}}$$

If, therefore, R_e is negligibly small, as in the case of a heavy copper wire, the current flowing through it will be n times as great as that which could be made to flow through it by a single cell. These considerations show that the rules which should govern the combination of cells are as follows:

When the external resistance is large in comparison with the internal resistance of a single cell, the cells should be connected in series.

When the external resistance is small in comparison with the internal resistance of a single cell, the cells should be connected in parallel.

The applications of Ohm's law to electric circuits are so numerous and varied that they will be more fully treated in the next book of this series on "Electrical Measurements and Applications of the Electric Current". In order better to understand Ohm's law as applied to what has preceded, however, a few illustrative examples, and examples for practice are given.

EXAMPLES FOR PRACTICE

1. A Daniell cell is found to send a current of 0.5 ampere through an ammeter of negligible resistance. What is the internal resistance of the cell?

$$I=\frac{E}{R_e+R_i} \text{ or } .5=\frac{1.08}{0+R_i}$$

$$R_i=\frac{1.08}{.5}=2.16$$

Therefore, the internal resistance of the cell is 2.16 ohms.

2. What is the resistance of a lamp filament if a p.d. of 110 volts applied to its terminals sends a current of 0.5 ampere through it?

$$I=\frac{\text{p.d.}}{R} \text{ or } .5=\frac{110}{R}$$

$$R=\frac{110}{.5}=220 \text{ ohms}$$

Therefore, the lamp filament has a resistance of 220 ohms.

3. How much current will 20 Daniell cells in series send through a telegraph line having a resistance of 300 ohms, if the internal resistance of each cell is 4 ohms, and its e.m.f. 1.08 volts?

$$I=\frac{NE}{R_e+NR_i}=\frac{20\times1.08}{300+(20\times4)}$$

$$=\frac{21.6}{380}=.057 \text{ (approx.)}$$

Therefore, the current is .057 ampere.

4. What current will the 20 cells of Example 3 arranged in parallel send through the same telegraph line?

$$I=\frac{E}{R_e+\frac{R_i}{N}}=\frac{1.08}{300+\frac{4}{20}}$$

$$=\frac{1.08}{300.2}=.0036 \text{ (approx.)}$$

Therefore, the current is .0036 ampere.

5. What current will one of the Daniell cells of Example 3 send through the same telegraph line?

$$I=\frac{E}{R_e+R_i}=\frac{1.08}{300+4}$$

$$=\frac{1.08}{304}=.0036 \text{ (approx.)}$$

Therefore, the current is .0036 ampere.

6. Would it be possible to increase appreciably the current through the above telegraph line by connecting a very large number of Daniell cells in parallel? (Compare answers to Examples 4 and 5.)

7. When the external resistance is large in comparison with that of a single cell, what method of connecting the cells yields the largest current?

8. What current will the 20 cells of Example 3, arranged in parallel, send through an ammeter which has a resistance of 0.05

ohm? What current would the 20 cells in series send through the same ammeter? What current would a single cell send through the same ammeter?

9. From the answers obtained in the different parts of Example 8, decide whether, in order to yield as large a current as possible, cells should be connected in series or in parallel, when the external resistance is small in comparison with that of a single cell.

10. Why is a Daniell cell a very poor one to use on a doorbell circuit? Why is a Leclanché a good one for the same purpose?

11. Where and for what purpose have you seen dry cells used?

12. A given new dry cell has an e.m.f. of 1.5 volts and gives a current of 25 amperes when short-circuited with a coarse copper wire. What is its internal resistance? After this cell was used on a doorbell circuit for one year, it gave only 1 ampere; what was its internal resistance then?

13. Why will two dry cells run a toy motor very rapidly, while three gravity cells having the same voltage as the two dry cells may not even move it?

THE MOST POWERFUL ELECTROMAGNET IN THE WORLD. CONSTRUCTED FOR THE NATIONAL BUREAU OF

ELEMENTS OF ELECTRICITY AND MAGNETISM

PART II

MEASUREMENT AND APPLICATIONS OF THE ELECTRIC CURRENT

OHM'S LAW

69. Introduction. A thorough mastery of Ohm's law bears the same relationship to the electrical engineer and his work, that the skillful manipulation of the hammer, saw, and square, does to the carpenter and his work. For this reason the quantities, *electromotive force*, *current*, and *resistance*, which were explained in Part I, are briefly reviewed here. Again, Ohm's law is used so extensively in the development of methods for comparing and measuring electromotive forces, currents, and resistances, that many of the applications of the law have been given in the following pages, with the double purpose in view, *first*, of acquiring ability to apply Ohm's law to practical problems; and, *second*, of familiarizing the student with the law in order that he may readily understand its application in developing the methods used in comparing electromotive forces with the potentiometer, in comparing resistances with the Wheatstone bridge, and so on.

70. Electromotive Force. When a difference of electrical potential exists between two points, there is said to exist an *electromotive force*, or a tendency to cause a current to flow from one point to the other. In the voltaic cell, one plate is at a different potential from the other, which gives rise to an electromotive force between them. In the induction coil, also, an electromotive force is created in the secondary circuit caused by the action of the primary. This electromotive force is analogous to the *pressure* caused by a difference in level of two bodies of water connected by a pipe. The pressure tends to force the water through the pipe, and the electromotive force tends to cause an electric current to flow.

The terms potential difference and electromotive force are commonly used with the same meaning, but strictly speaking the electromotive force gives rise to the potential difference. Electromotive force is commonly designated by the letters *e.m.f.* or simply *E.* It is also referred to as *pressure* or *voltage.*

71. Current. A current of electricity flows when two points at a difference of potential are connected by a wire, or when the circuit is otherwise completed. Similarly, water flows to a lower level when a path is provided. In either case, the flow can take place only when the path exists. Hence, to produce a current, it is necessary to have an electromotive force and a closed circuit. The current continues to flow only as long as the electromotive force and closed circuit exist.

The strength of a current in a conductor is defined as the quantity of electricity which passes any point in the circuit in a unit of time.

Current is sometimes designated by the letter *C*, but the letter *I* will be used for current throughout this and following sections. The latter symbol was recommended by the International Electrical Congress held at Chicago in 1893, and has since been universally adopted.

72. Resistance. Resistance is that property of matter by virtue of which bodies oppose or resist the free flow of electricity. Water passes with difficulty through a small pipe of great length or through a pipe filled with stones or sand, but very readily through a large clear pipe of short length. Likewise a small wire of considerable length and made of poor conducting material offers great resistance to the passage of electricity, but a good conductor of short length and large cross-section offers very little resistance.

Resistance is designated by the letter *R.*

Volt, Ampere, and Ohm. The *volt* is the practical unit of electromotive force.

The *ampere* is the practical unit of current.

The *ohm* is the practical unit of resistance. The *microhm* is one-millionth of an ohm and the *megohm* is one million ohms.

The standard values of the above units were very accurately determined by the International Electrical Congress in 1893, and are as follows:

The International ohm, or true ohm is, as nearly as known, the resistance of a uniform column of mercury 106.3 centimeters long and 14.4521 grams in mass, at the temperature of melting ice.

The ampere is the strength of current, which, when passed through a solution of silver nitrate under suitable conditions, deposits silver at the rate of .001118 gram per second. Current strength may be very accurately determined by electrolysis, and it is used therefore in determining the standard unit.

The volt is equal to the e.m.f. which, when applied to a conductor having a resistance of one ohm, will produce in it a current of one ampere. One volt equals $\frac{10000}{10183}$ of the e.m.f. of a standard Weston cell at 20° C.

73. Ohm's Law Defined. *The current is directly proportional to the electromotive force, and inversely proportional to the resistance.*

That is, if the electromotive force applied to a circuit is increased, the current will be increased in the same proportion, and if the resistance of a circuit is increased, then the current will be decreased proportionally. Likewise a decrease in the electromotive force causes a proportional decrease in current and a decrease in resistance causes a proportional increase in current. The current depends only upon the electromotive force and resistance and in the manner expressed by the above simple law. The law may be expressed algebraically as follows:

$$\text{current } \textit{varies as } \frac{\text{electromotive force}}{\text{resistance}}$$

The units of these quantities, the ampere, the volt, and the ohm, have been so chosen that an electromotive force of 1 volt applied to a resistance of 1 ohm, causes 1 ampere of current to flow. Ohm's law may therefore be expressed by the following equation

$$\text{amperes} = \frac{\text{volts}}{\text{ohms}}$$

or

$$I = \frac{E}{R}$$

where I is the current in amperes, E the electromotive force in volts, and R the resistance in ohms.

It is therefore evident, that if the electromotive force and the resistance are known, the current may be found, or if any two of the three quantities are known, the third may be found. If the current and the resistance are known, the electromotive force may be found from the formula

$$E = RI$$

and if the current and the electromotive force are known, the resistance may be found from the formula

$$R = \frac{E}{I}$$

74. Simple Applications. The following examples are given to illustrate the simplest applications of Ohm's law.

Examples. 1. If the e.m.f. applied to a circuit is 4 volts and its resistance is 2 ohms, what current will flow?

By the formula for current

$$I = \frac{E}{R} = \frac{4}{2} = 2 \text{ amperes} \qquad \text{Ans. 2 amperes}$$

2. What voltage is necessary to cause a current of 23 amperes to flow through a resistance of 820 ohms?

By the formula for e.m.f.

$$E = RI = 820 \times 23 = 18{,}860 \text{ volts}$$

Ans. 18,860 volts

3. The e.m.f. applied to a circuit is 110 volts, and it is desired to obtain a current of .6 ampere. What should be the resistance of the circuit?

By the formula for resistance

$$R = \frac{E}{I} = \frac{110}{.6} = 183 + \text{ohms} \qquad \text{Ans. 183 ohms}$$

75. Series Circuits. A circuit made up of several parts all joined in series with each other, is called a series circuit and the resistance of the entire circuit is of course the sum of the separate resistances. In calculating the current in such a circuit, the total resistance must first be obtained, and the current may then be found by dividing the applied or total e.m.f. by the total resistance. This is expressed by the formula

$$I=\frac{E}{r_1+r_2+r_3+\text{etc.}}$$

Examples. 1.. Three resistance coils are connected in series with each other and have a resistance of 8, 4, and 17 ohms, respectively. What current will flow if the e.m.f. of the circuit is 54 volts?

By the preceding formula

$$I=\frac{E}{r_1+r_2+r_3}=\frac{54}{8+4+17}=\frac{54}{29}=1.8+ \text{ amperes}$$

Ans. 1.8 amperes

2. Six arc lamps, each having a resistance of 5 ohms, were connected in series with each other, and the resistance of the connecting wires and other apparatus was 3.7 ohms. What must have been the pressure of the circuit to give a desired current of 9.6 amperes?

The total resistance of the circuit was $R=(6\times5)+3.7=33.7$ ohms and the current was to be $I=9.6$ amperes. Hence by the formula for e.m.f.

$$E=RI=33.7\times9.6=323+ \text{ volts} \qquad \text{Ans. 323 volts}$$

3. The current passing in a certain circuit was 12 amperes and the e.m.f. was 743 volts. The circuit was made up of 4 sections all connected in series, and the resistance of three sections was 16, 9, and 26 ohms, respectively. What was the resistance of the fourth section?

Let x = the resistance of the fourth section, then $R=16+9+26+x=51+x$, $I=12$, and $E=743$. By the formula for resistance

$$R=\frac{E}{I}, \text{ or } 51+x=\frac{743}{12}=61.9 \text{ ohms (approx.)}$$

If $51+x=61.9$ we have, by transposing 51 to the other side of the equation

$$x=61.9-51=10.9 \text{ ohms} \qquad \text{Ans. 10.9 ohms}$$

4. A current of 54 amperes flowed through a circuit when the e.m.f. was 220 volts. What resistance should be added in series with the circuit to reduce the current to 19 amperes?

The resistance in the first case was

$$R=\frac{220}{54}=4.07 \text{ ohms (approx.)}$$

The resistance in the second must be

$$R=\frac{220}{19}=11.58 \text{ ohms (approx.)}$$

The required resistance to insert in the circuit is the difference of these two resistances, or $11.58-4.07=7.51$ ohms.

Ans. 7.51 ohms

76. Fall of Potential in a Circuit. Fig. 81 illustrates a series circuit in which the resistances *A*, *B*, *C*, *D*, and *E* are connected in series with each other and with the source of electricity. If the e.m.f. is known, the current may be found by dividing the e.m.f. by

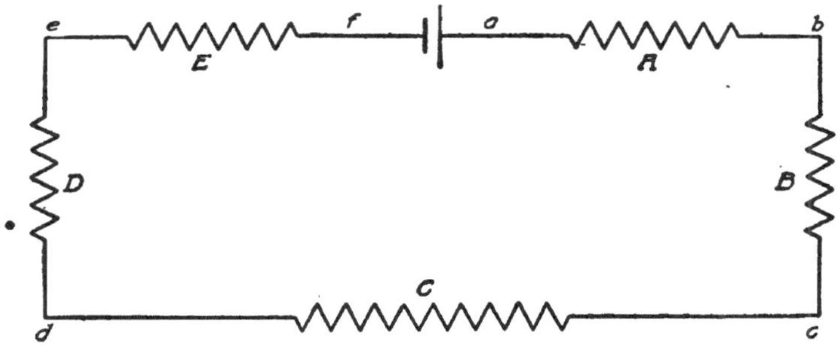

Fig. 81. Diagram of Series Circuit Illustrating Fall Potential

the sum of all the resistances. Ohm's law may, however, be applied to any part of a circuit separately, as well as to the complete circuit. Suppose the resistances of *A*, *B*, *C*, *D*, and *E* are 4, 3, 6, 3, and 4 ohms, respectively, and assume that the source has no resistance. Suppose the current flowing to be 12 amperes. The e.m.f. necessary to force a current of 12 amperes through the resistance *A* of 4 ohms is, by applying Ohm's law, equal to $E=RI=4\times12=48$ volts. Hence, between the points *a* and *b* outside of the resistance *A*, there must be a difference of potential of 48 volts to force the current through this resistance. Also, to force the same current through *B*, the voltage necessary is $3\times12=36$. Similarly, for each part *C*, *D*, and *E*, there are required 72, 36, and 48 volts, respectively.

As 48 volts are necessary for part *A*, and 36 volts for part *B*, it is evident that to force the current through both parts a difference of potential of 48+36=84 volts is required; that is, the voltage between the points *a* and *c* must be 84 volts. For the three parts *A*, *B*, and *C*, 48+36+72=156 volts are necessary, and for the entire circuit, 240 volts must be applied to give the current of 12 amperes. From the above it is evident that there is a gradual fall of potential throughout the circuit, and if the voltage between any two points of the circuit be measured, the e.m.f. obtained would depend upon the resistance included between these two points. For example, the voltage between points *b* and *d* would be found to be 72+36=108 volts, or between *d* and *e* 36 volts, etc. From the preceding it is apparent that the fall of potential in a part of a circuit is equal to the current multiplied by the resistance of that part.

This gradual fall of potential, or *drop* as it is commonly called, throughout a circuit, enters into the calculations for the size of conductors or mains supplying current to distant points. The resistances of the conductors cause a certain drop in transmitting the current, depending upon their sizes and lengths, and it is therefore necessary that the voltage of machines at the supply station shall be great enough to give the voltage necessary at the receiving stations as well as the additional voltage lost in the conducting mains.

For example, in Fig. 81 the voltage necessary between the points *e* and *b* is 144 volts, but to give this voltage, the source must supply in addition the voltage lost in parts *A* and *E*, which equals 96 volts.

Examples. 1. The voltage required by 17 arc lamps connected in series is 782 volts and the current is 6.6 amperes. The resistance of the connecting wires is 7 ohms. What must be the e.m.f. applied to the circuit.

The drop in the connecting wires is $E=RI=7\times 6.6=46.2$ volts. The e.m.f. necessary is therefore 782+46.2=828.2 volts.

Ans. 828.2 volts

2. A source of e.m.f. supplies 114 volts to a circuit made up of incandescent lamps and conducting wires. The lamps require a voltage of 110 at their terminals, and take a current of 12 amperes. What should be the resistance of the conducting wires in order that the lamps may receive the necessary voltage?

The allowable drop in the conducting wires is 114−110=4 volts. The current to pass through the wires is 12 amperes. Hence the resistance must be

$$R=\frac{E}{I}=\frac{4}{12}=.33 \text{ ohm}$$

Ans. .33 ohm

77. Divided Circuits. When a circuit divides into two or more parts, it is called a *divided* circuit and each part will transmit a portion of the current.

Such a circuit is illustrated in Fig. 82, the two branches being represented by *b* and *c*. The current passes from the positive pole of the battery through *a* and then divides; part of the current passing through *b* and part through *c*. The current then unites and passes through *d* to the negative pole of the battery. The part *c* may be considered as the main part of the circuit and *b* as a by-pass about it. A branch which serves as a by-pass to another circuit is called a *shunt* circuit, and the two branches are said to be connected in *parallel*.

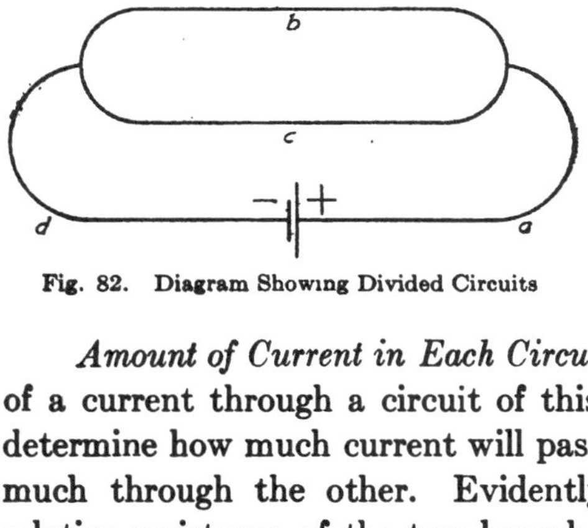

Fig. 82. Diagram Showing Divided Circuits

Amount of Current in Each Circuit. In considering the passage of a current through a circuit of this sort, it may be necessary to determine how much current will pass through one branch and how much through the other. Evidently this will depend upon the relative resistance of the two branches, and more current will pass through the branch offering the lesser resistance than through the one offering the greater. If the two parts have equal resistances, then one-half of the total current will pass through each branch. If one branch has twice the resistance of the other, then only one-half as much of the total current will pass through that branch as through the other; that is, $\frac{1}{3}$ of the total current will pass through the first branch and the remaining $\frac{2}{3}$ will pass through the second.

The relative strength of current in the two branches will be inversely proportional to their resistances, or directly proportional to their conductances.

Suppose the resistance of one branch of a divided circuit is r_1, Fig. 83, and that of the other is r_2. Then by the preceding law

$$\text{current in } r_1 : \text{current in } r_2 : : r_2 : r_1$$

and

$$\text{current in } r_1 : \text{total current} : : r_2 : r_1+r_2$$

and

$$\text{current in } r_2 : \text{total current} : : r_1 : r_1+r_2$$

Let I represent the total current, i_1 the current through the resistance r_1, and i_2 the current through the resistance r_2. Then

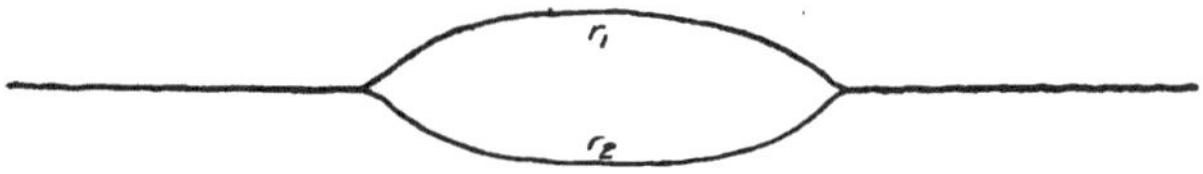

Fig. 83. Diagram of Two Resistances in Parallel

the two preceding proportions are expressed by the following formulas

$$i_1 = \frac{Ir_2}{r_1+r_2} \text{ and } i_2 = \frac{Ir_1}{r_1+r_2}$$

Example. The total current passing in a circuit is 24 amperes. The circuit divides into two branches having resistances of 5 and 7 ohms, respectively. What is the current in each branch?

In this case $I=24$, $r_1=5$, and $r_2=7$. Substituting these values in the above formulas, we have

$$i_1 = \frac{Ir_2}{r_1+r_2} = \frac{24\times 7}{5+7} = 14 \text{ amperes}$$

and

$$i_2 = \frac{Ir_1}{r_1+r_2} = \frac{24\times 5}{5+7} = 10 \text{ amperes}$$

Ans. $\begin{cases}\text{In 5-ohm branch, 14 amperes}\\ \text{In 7-ohm branch, 10 amperes}\end{cases}$

78. Joint Resistance of Divided Circuits. As a divided circuit offers two paths to the current, it follows that the joint resistance of the two branches will be less than the resistance of either branch alone. The ability of a circuit to conduct electricity is represented by its conductance, which is the reciprocal of resistance; and the conductance of a divided circuit is equal to the sum of the conductances of its parts.

For example, in Fig. 83, the conductance of the upper branch equals $\frac{1}{r_1}$ and that of the lower branch equals $\frac{1}{r_2}$. If R represents the joint resistance of the two parts, then the joint conductance equals

$$\frac{1}{R}=\frac{1}{r_1}+\frac{1}{r_2}=\frac{r_1+r_2}{r_1 r_2}$$

Having thus obtained the joint conductance, the joint resistance is found by taking the reciprocal of the conductance, that is

$$R=\frac{r_1 r_2}{r_1+r_2}$$

This formula may be stated as follows:

The joint resistance of a divided circuit is equal to the of the two separate resistances divided by their sum.

Fig. 84. Diagram of Three Resistances in Parallel

For example, suppose the resistance of each branch to be 2 ohms. The conductance of the circuit will be

$$\frac{1}{R}=\frac{1}{2}+\frac{1}{2}=1, \text{ and hence } R=1 \text{ ohm}$$

Also by the preceding formula

$$R=\frac{2\times 2}{2+2}=1 \text{ ohm}$$

The resistance of a divided circuit in which each branch has a resistance of 2 ohms is therefore 1 ohm.

Example. The resistances of two separate conductors are 3 and 7 ohms, respectively. What would be their joint resistance if connected in parallel?

In this case $r_1=3$ and $r_2=7$. Hence, by the formula

$$R=\frac{3\times 7}{3+7}=2.1 \text{ ohms} \qquad \text{Ans. 2.1 ohms}$$

Calculating Joint Resistance of Three Conductors. Suppose, as illustrated in Fig. 84, the conductors having resistances equal to r_1, r_2, and r_3, respectively, are connected in parallel. The joint total conductance will then be equal to

$$\frac{1}{R}=\frac{1}{r_1}+\frac{1}{r_2}+\frac{1}{r_3}=\frac{r_2r_3+r_1r_3+r_1r_2}{r_1r_2r_3}$$

and as the joint resistance is the reciprocal of the joint conductance, the joint resistance R of the three branches is expressed by the formula

$$R=\frac{r_1r_2r_3}{r_2r_3+r_1r_3+r_1r_2}$$

Example. What is the joint resistance when connected in parallel, of three wires whose resistances are 41, 52, and 29 ohms, respectively?

In this case $r_1=41$, $r_2=52$, and $r_3=29$. Hence, by the preceding formula

$$R=\frac{41\times52\times29}{52\times29+41\times29+41\times52}=12.8+\text{ ohms}$$

Ans. 12.8+ ohms

Joint Resistance of Any Number of Conductors. In general, for any number of conductors connected in parallel, the joint resistance is found by taking the reciprocal of the sum of the reciprocals of the separate resistances.

Example. A circuit is made up of five wires connected in parallel, and their separate resistances are, respectively, 12, 21, 28, 8, and 42 ohms. What is the joint resistance?

The sum of the conductances is

$$\frac{1}{12}+\frac{1}{21}+\frac{1}{28}+\frac{1}{8}+\frac{1}{42}=\frac{53}{168}$$

Hence, the joint resistance equals

$$R=\frac{168}{53}=3.1+\text{ ohms}$$

Ans. 3.1+ ohms

Calculating Current in Each Conductor. If the resistance of each branch is known, and also the potential difference between the points of union, then the current in each branch may be found by applying Ohm's law to each branch separately. For example, if this potential difference were 96 volts, and the separate resistances of the 4 branches were 8, 24, 3, and 48 ohms, respectively, then the current in the respective branches would be 12, 4, 32, and 2 amperes, respectively.

If the current in each branch is known, and also the potential difference between the points of union, then the resistance of each branch may likewise be found from Ohm's law.

In general, if several resistances are connected in parallel, the following method will be found very practical and is just as accurate as the values of the individual resistances:

Example. Four incandescent lamps, having resistances of 220, 109, 605, and 297 ohms, respectively, are connected in parallel. What is their joint resistance?

In this case $r_1=220$, $r_2=109$, $r_3=605$, $r_4=297$.

The sum of the conductances $\frac{1}{R}$ is

$$\frac{1}{R}=\frac{1}{220}+\frac{1}{109}+\frac{1}{605}+\frac{1}{297}$$
$$=.00455+.00917+.00166+.00336$$
$$=.01874$$
$$R=\frac{1}{.01874}=53.4 \text{ ohms}$$

Ans. 53.4 ohms (approx.)

EXAMPLES FOR PRACTICE

1. Two conductors having resistances of 71 and 19 ohms, respectively, are connected in parallel, and the total current passing through the circuit is 37 amperes. What is the current in the conductor whose resistance is 71 ohms? Ans. 7.8+ amperes

2. What is the joint resistance of two wires connected in parallel if their separate resistances are 2 and 8 ohms, respectively? Ans. 1.6 ohms

3. What is the joint resistance of three wires when connected in parallel, whose separate resistances are 5, 7, and 9 ohms, respectively? Ans. 2.2+ ohms

4. Three wires, the respective resistances of which are 8, 10, and 20 ohms, are joined in parallel. What is their joint resistance? Ans. 3.6+ ohms

5. Four wires are joined in parallel, and their separate resistances are 2, 4, 6, and 9 ohms, respectively. What is the joint resistance of the conductor thus formed? Ans. .97+ ohm

6. Four lamps are joined in parallel, and their separate resistances are 209, 107, 583, and 275 ohms, respectively. What is their joint resistance? Ans. 51.3+ ohms

79. Battery Circuits. *Single Cell in Series with Resistance.* Fig. 85 illustrates a simple circuit having a single cell C connected in series with a resistance. This is the customary manner of representing a cell, the short heavy line representing the zinc and the long light line representing the copper or carbon plate. In determining the amount of current which will flow in such a circuit, the total resistance of the circuit must be considered. This is made up of the external resistance R and the internal resistance r or the resistance of the cell itself. If E represents the total e.m.f. of the cell, then the current I which will flow is expressed by the formula

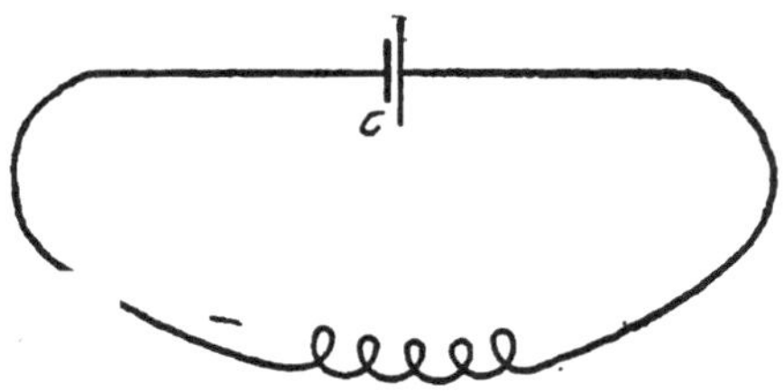

Fig. 85. Simple Battery Circuit

$$I = \frac{E}{R+r}$$

It has been shown that whenever a current passes through any resistance, there is always a certain *drop* or fall of potential. The total e.m.f. above referred to expresses the total potential difference between the plates of the cell and is the e.m.f. of the cell on *open* circuit. When the current flows, however, there is a fall of potential or loss of voltage within the cell itself, and hence the e.m.f. of the cell on closed circuit is less than that on open circuit. That is, if the voltage be measured when the cell is supplying current, it will be found to be less than when the voltage is measured on open circuit, or when the cell is supplying no current. The voltage on closed circuit is that available for the external circuit, and is therefore called the *external* or *available* voltage or e.m.f.

The external e.m.f. depends of course upon the strength of current the cell is supplying, and may be calculated as follows:

If the current passing is I, and the resistance of the cell is r, then, according to Ohm's law, the voltage lost in the cell equals

rI. If E represents the total e.m.f. of the cell, and E_1 the external e.m.f. then

$$E_1 = E - rI$$

The e.m.f. of a cell is understood to be the total e.m.f. unless otherwise stated.

Two or More Cells in Series. When two or more cells are interconnected they are said to form a *battery*.

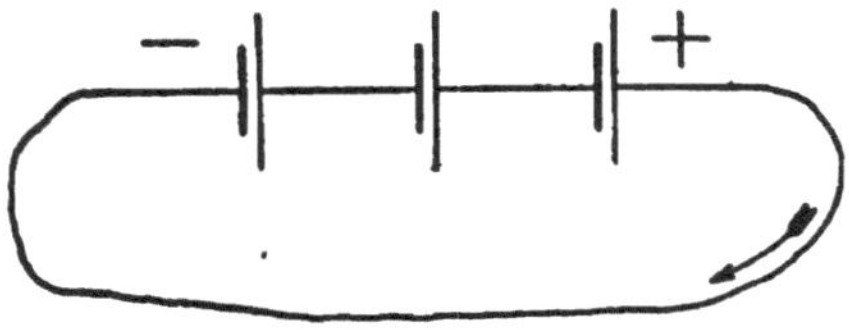

Fig. 86. Diagram of Three Cells in Series

Fig. 86 illustrates three cells connected in series with each other and with the external circuit. That is, the positive terminal of one cell is connected to the negative of the next, and the positive of that cell to the negative of the adjacent, etc. By this method of connecting, the e.m.f. of each cell is added to that of the others, so that the total e.m.f. of the circuit is three times that of a single cell. If one of the cells were connected so that its e.m.f. opposed that of the other two, it would offset the e.m.f. of one of the cells and the resultant e.m.f. would be that of a single cell. The connecting of cells in series as in Fig. 86 not only increases the e.m.f. of the circuit but also increases the internal resistance, the resistance of each cell being added to that of the others. If E equals the e.m.f. of each cell, r the internal resistance of each, and R the external resistance, then the current that will flow is expressed by the formula

$$I = \frac{3E}{R+3r}$$

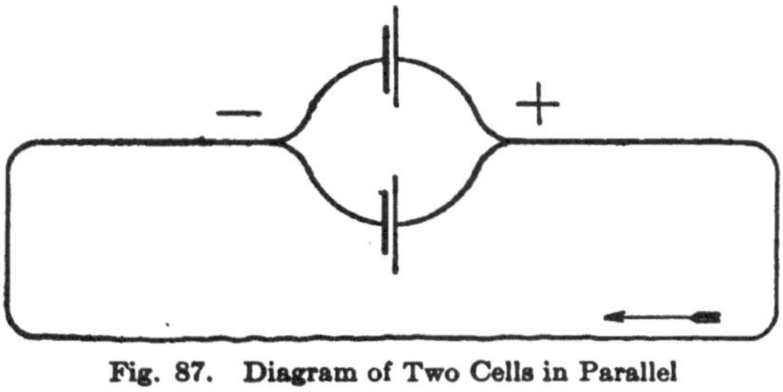

Fig. 87. Diagram of Two Cells in Parallel

or for n cells connected in series the formula for current is

$$I = \frac{nE}{R+nr}$$

Two or More Cells in Parallel. Fig. 87 illustrates two cells connected in parallel, and supplying current to an external circuit. Here the two positive terminals are connected with each other and the two negative are also connected with each other. The e.m.f. supplied to the circuit is equal to that of a single cell only. In fact, connecting cells in parallel is equivalent to enlarging the plates, and the only effect is to decrease the internal resistance. It is evident that coupling two cells in parallel affords two paths for the current and so decreases the resistance of the two cells to one-half that of a single cell. The formula expressing the current that would flow in the external circuit with two cells in parallel is, therefore,

$$I=\frac{E}{R+\frac{r}{2}}$$

or for n cells connected in parallel, the formula for current is

$$I=\frac{E}{R+\frac{r}{n}}$$

Combination of Cells in Series and in Parallel. Fig. 88 represents a combination of the series and the parallel methods of connecting, four files of cells being joined in parallel, each file having four cells connected in series. The e.m.f. of each file and consequently of the circuit is $4E$. The resistance of each file is $4r$, and that of all the files $\frac{4r}{4}$. Hence, the formula for current is

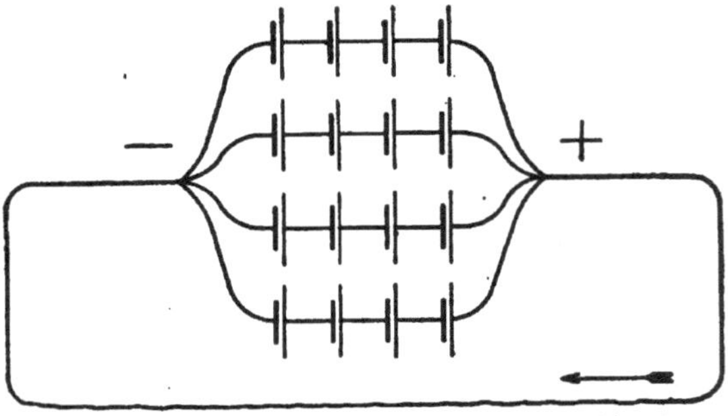

Fig. 88. Group of Cells in Series-Parallel Circuit

$$I=\frac{4E}{R+\frac{4r}{4}}$$

If there were m files connected in parallel, and n cells were connected in series in each file, the formula expressing the current in the external circuit would be

$$I=\frac{nE}{R+\frac{nr}{m}}$$

where E is the e.m.f. of each cell, R the external resistance, and r the internal resistance of each cell.

Method of Connecting Cells for Giving Best Results. The most advantageous method of connecting cells depends upon the results desired, the resistance of the cell, and the external resistance. Suppose it is desired to pass a current through an external resistance of 2 ohms, and that Daniell cells are to be used each having an e.m.f. of 1 volt and an internal resistance of 3 ohms. With one cell only in the circuit, the current will be

$$\frac{E}{R+r}=\frac{1}{2+3}=.2 \text{ ampere}$$

and with 5 cells all in series, the current will be

$$\frac{5E}{R+5r}=\frac{5}{2+15}=.3 \text{ ampere (approx.)}$$

Therefore, with 5 cells in series, the current is only .1 ampere greater than that with a single cell, and with 100 cells in series the current is only

$$\frac{100\,E}{R+100r}=\frac{100}{2+300}=.33 \text{ ampere}$$

Hence, with a comparatively low external resistance, there is but little gain in current strength by the addition of cells in series. This is due to the fact that, although the e.m.f. is increased by 1 volt for each cell, the resistance is increased by 3 ohms.

Now suppose 5 Daniell cells to be connected in parallel with the external circuit of 2 ohms. The e.m.f. of the circuit will then be that of a single cell, and the current will be

$$\frac{E}{R+\frac{r}{5}}=\frac{1}{2+\frac{3}{5}}=.4 \text{ ampere (approx.)}$$

and with 100 cells connected in parallel the current will be

$$\frac{E}{R+\frac{r}{100}}=\frac{1}{2+\frac{3}{100}}=.5 \text{ ampere (approx.)}$$

A larger current is therefore obtained in this case by connecting the cells in parallel than by connecting them in series.

With a large external resistance, on the other hand, a larger current is obtained by connecting the cells in series. For example, suppose the external resistance to be 500 ohms. One cell will then give a current of .00198+ ampere, and 5 cells in series will give about .0097 ampere, whereas 100 cells will give .125 ampere. With 5 cells connected in parallel, the current will be .00199+ ampere, and with 100 cells the current will amount to approximately .002 ampere. With an external resistance of 500 ohms, there is practically no advantage in connecting the cells in parallel. The only effect of the latter method is to decrease the internal resistance which is almost negligible in comparison with the external resistance.

It may be shown mathematically that for a given external resistance and a given number of cells, the largest current is obtained when the internal resistance is *equal* to the external resistance. In order to obtain this result, the values of m and n in the formula on page 81, should be so chosen that $\frac{nr}{m}$ equals R. This arrangement, although giving the largest current strength, is not the most economical. With the internal resistance equal to the external, there is just as much energy used up in the battery itself as is expended usefully in the external circuit.

In order to obtain the most economical arrangement, the internal resistance should be made as small as possible, that is, all the cells should be connected in parallel. The loss of power in the battery is then the smallest amount possible.

In order to obtain the quickest action of the current, the cells should be connected in series. When the external circuit possesses considerable self-induction, as is the case when electromagnets are connected in the circuit, the action of the current is retarded, since, as will be shown later, the self-induction acts like a counter e.m.f. in the circuit. This retardation must be overcome by having a high e.m.f., which is obtained by connecting the cells in series.

Examples. 1. Sixteen cells, each having an internal resistance of .1 ohm are to be connected with a circuit whose resistance is .4 ohm. How should the cells be connected to obtain the greatest current?

Here the external resistance R equals .4 ohm and the resistance r of each cell equals .1 ohm. For maximum current

$$\frac{nr}{m} = R, \text{ or } \frac{.1n}{m} = .4$$

Therefore

$$.1n = .4m, \text{ or } n = 4m$$

and as $mn=16$, the only values of m and n which will be true for both of these equations are $n=8$ and $m=2$. Hence there must be 2 files of cells, with 8 cells in series in each file.

Ans. 2 files, 8 cells in each

2. The external resistance in a circuit is 4 ohms. The cells used each have an e.m.f. of 1.2 volts and an internal resistance of 3.8 ohms. If 20 cells were used, which method of connecting would supply the larger current, 5 files with 4 cells in series, or 4 files with 5 cells in series?

(a) Applying the formula on page 81, we have $R=4$, $E=1.2$, $r=3.8$; and with 5 files and 4 in series, $n=4$ and $m=5$. Hence, the current is

$$\frac{nE}{R+\frac{nr}{m}}=\frac{4\times1.2}{4+\frac{4\times3.8}{5}}=.681+\text{ ampere}$$

With 4 files and 5 cells in series. $n=5$ and $m=4$. Hence the current is

$$\frac{5\times1.2}{4+\frac{5\times3.8}{4}}=.685+\text{ ampere}$$

The larger current is therefore supplied by having 4 files with 5 cells in series. Ans. 4 files with 5 cells in series

(b) The maximum current is supplied when the internal resistance equals the external resistance, or when

$$\frac{nr}{m}=R$$

With 5 files and 4 cells in series

$$\frac{nr}{m}=\frac{4\times3.8}{5}=3.04\text{ ohms}$$

and with 4 files and 5 cells in series

$$\frac{nr}{m}=\frac{5\times3.8}{4}=4.75\text{ ohms}$$

The latter value is nearer to 4 ohms, which is the external resistance, than is 3.04. Hence, the larger current will be supplied with 4 files and 5 cells in series. Ans. 4 files with 5 cells in series

3. It is desired to pass a current of .025 ampere through an external resistance of 921 ohms. The cells are to be connected in series and each has an e.m.f. of .8 volt and an internal resistance of 1.3 ohms. What number of cells must be used?

From page 80, the general formula for cells in series is

$$I = \frac{nE}{R+nr}$$

and in this case $I=.025$; $E=.8$; $R=921$; and $r=1.3$. Substituting these values gives

$$.025 = \frac{n \times .8}{921 + n1.3}$$

Multiplying by $921+1.3n$ gives

$$23.025 + .0325n = .8n$$

Transposing .0325 n gives

$$.8n - .0325n = 23.025$$

$$.7675n = 23.025$$

$$n = 30$$

Ans. 30 cells

EXAMPLES FOR PRACTICE

1. Ten cells in series have an e.m.f. of 1 volt each and an internal resistance of .2 ohm. The external resistance is 3 ohms. What is the current? Ans. 2 amperes

2. Six cells, each of which has an e.m.f. of 1.2 volts and a resistance of 2 ohms, are connected in parallel. With an external resistance of 10 ohms, what is the current? Ans. .116+ ampere

3. What is the current supplied by the same cells if they are joined in series and the external resistance is 20 ohms?

Ans. .225 ampere

4. A single cell whose e.m.f. on open circuit is 1.41 volts and whose internal resistance is .5 ohm is supplying a current of .3 ampere. Determine the value of the available e.m.f. of the cell?

Ans. 1.26 volts

5. What would be the available e.m.f. with 8 of the cells referred to in example 4, when connected in series and supplying the same current? Ans. 10.08 volts

6. Eight Daniell cells (e.m.f. = 1.05, resistance = 2.5 ohms each) are joined in series. Three wires A, B, and C of 9, 36, and

72 ohms resistance, respectively, are arranged to be connected to the poles of the battery. Find the current when each wire is inserted separately, and when all three wires are connected in parallel.

Ans. Through *A*, .29 ampere approx.; through *B*, .15 ampere; through *C*, .091+ ampere; and through all three, .31+ ampere.

7. A battery of 28 bunsen cells (e.m.f. = 1.8, resistance .1 ohm each) are to supply current to a circuit having an external resistance of 30 ohms. Find the current (a) when all the cells are joined in series, (b) when all the cells are in parallel, (c) when there are 2 files each having 14 cells in series, (d) when there are 7 files each having 4 cells in series.

Ans. (a) 1.53+; (b) .06 nearly; (c) .82+; (d) .23+ ampere.

ELECTROLYSIS—MEASUREMENT OF CURRENT STRENGTH

80. Electric Current Conducted through Liquid. The way in which a current is conducted through a liquid is very beautifully shown by the following experiment: Prepare an infusion of purple cabbage either by steeping the cabbage thoroughly in water, or by grinding the cabbage and then pressing out the juice and adding water. To a portion of this infusion add a very small quantity of caustic soda (NaOH), or any other alkali, and the infusion will be turned green. To another portion of the infusion, add a few drops of sulphuric acid (H_2SO_4), or any other acid, and the infusion will be turned red. We now know that an alkali will turn the infusion green and that an acid will turn it red, so that we may use this fact as a test for an alkali or an acid. Now take a third portion of the infusion and pour enough of it into a solution of sodium sulphate (Na_2SO_4) to give it a decided purple color. Place this solution in a U or a V tube, Fig. 89, insert the platinum electrodes *A* and *B* and send a current from three or four dry cells connected in series through it. The solution near the pole at which the current enters the liquid, called the *anode*, will presently be seen to turn red; while that

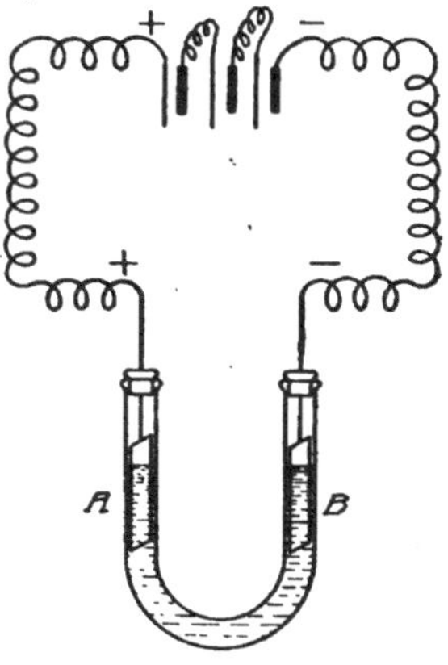

Fig. 89. Diagram of Electrolytic Cell

near the pole at which the current leaves the liquid, called the *cathode,* will turn green.*

These facts are explained by the theory of dissociation, see Section 62. Here the Na_2SO_4 when it goes into solution dissociates, i.e., it breaks up into negatively charged SO_4 ions and positively charged Na ions. Hence, as soon as the platinum plates *A* and *B* became positively and negatively charged, respectively, by being attached to the + and − plates of the battery, the negative SO_4 ions were at once attracted toward *A* and the positive NA ions toward *B*. When they reached the plates, they gave up their charges to them, and were then in condition to attack the water of the solution, forming H_2SO_4 at one plate and sodium hydroxide (NaOH) at the other.

All conduction in liquids, molten metals excepted, is thought to be due to a mechanism similar to that described above, i.e., it is thought to consist in the migration through the liquid of a swarm of positively charged ions in the direction in which the electric current is supposed to flow and of a corresponding swarm of negatively charged ions in the other direction. As soon as the ions give up their charges to the plates, such charges are either deposited upon these plates or else act chemically either upon the solution or upon the plate, so as to form new compounds. Strong evidence for the correctness of this view as to the nature of conduction in liquids is found in the fact that *pure* liquids, such as carefully distilled water, alcohol, etc., do not conduct electricity, but are, in every case, rendered conductors by dissolving salts or acids in them; and again, by the further fact that whenever an electric current is passed through such a conducting liquid, the two constituents of the substance in solution are found to appear at the two plates.

All liquids which conduct in this manner are called *electrolytes,* and the process of separating the two constituents of the substance in solution by means of an electric current is called *electrclysis.*

81. Electrolysis of Water. If two platinum electrodes, sealed into bent glass tubes *e* and *f*, Fig. 90, are inserted into two inverted

*This test may be made equally well with litmus in place of the cabbage infusion, but in this case it must be made in two parts. In the first part, the solution is first turned red with a few drops of acid; the passage of the current then causes it to turn *blue* about the *cathode*. In the second part, the solution is first turned blue by an excess of alkali (NaOH); the passage of the current then causes it to turn red about the *anode*.

test tubes, the test tubes and the vessel in which they stand being filled with about one part of sulphuric acid to forty parts of water, and if then two or three dry cells are connected in series to the terminals *A* and *B*, gases will immediately be seen rising from the two platinum electrodes and collecting in the top of the tubes. One tube will be found to fill twice as rapidly as the other.* If a lighted match is held in front of the tube *H* after it has been completely filled and removed, it will be found that this gas burns with a blue flame showing that it is hydrogen. If the gas in the tube *O* is tested by introducing into it a glowing splinter, the splinter will take fire and burn vigorously, showing that this tube contains oxygen.

The water has been decomposed by the electric current into its two elements, hydrogen and oxygen. According to our theory this decomposition has been effected as follows: The positively charged hydrogen ions of the sulphuric acid (H_2SO_4) solution were driven by the electric forces to the negative electrode, where they gave up their charges and appeared at once as hydrogen gas; while the negative SO_4 ions migrated to the positive electrode, where they gave up their charges to it and then acted upon the water (H_2O), forming more sulphuric acid and liberating oxygen.

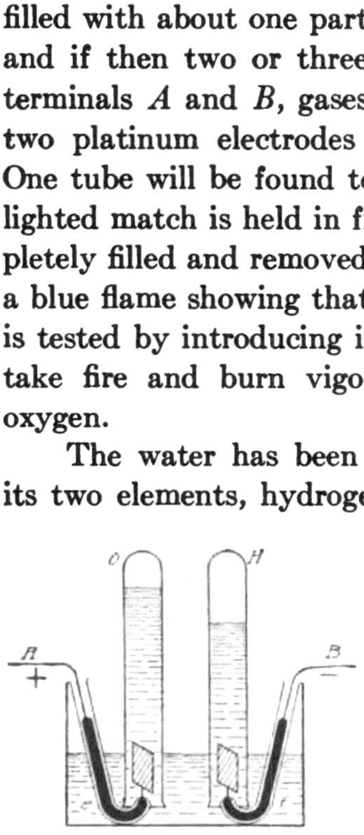

Fig. 90. Experiment Showing Electrolysis of Water

If the electrolysis of water is produced in the hydrogen voltameter of Fig. 91, it is found that 1 coulomb of electricity will liberate .0000104 gram of hydrogen, or .1156 cc. of hydrogen at 0° C. and 760 mm. pressure.

If we let V be the observed volume of hydrogen, V_0 the volume at zero, by the laws of Boyle and Charles (see Practical Physics)

$$V_0 = V \times \frac{P}{760} \times \frac{273}{273+t}$$

in which $P = \left(B + \frac{h}{12} - p\right)$. B is the barometer reading in mm., h is

*The oxygen will be absorbed by electrodes other than platinum, so that its volume will be much less than one-half that of the hydrogen.

the height in mm. of the acid solution in R above that in H, and p is the vapor pressure of saturated water vapor in mm. of mercury at the given temperature $t°$ C. Since the acid solution is about one-twelfth as heavy as mercury, h in the last equation is divided by 12 to reduce the pressure produced by this column of acid to mm. of mercury.

V_0 having been found, the formula for the number of coulombs of electricity which passes through the voltameter is

$$\text{coulombs} = \frac{V_0}{.1156}$$

and since

$$\text{coulombs} = \text{amperes} \times \text{seconds}$$

if we know how many seconds the current is on and the number of coulombs of electricity which pass through the voltameter, the current is at once given by

$$\text{amperes} = \frac{\text{coulombs}}{\text{seconds}}$$

If an ammeter is connected in series with the hydrogen voltameter, and the reading kept constant by means of a control resistance, Fig. 96, the ammeter will read the same as the amperage computed in the manner above, if the ammeter is correct and the experiment is carefully performed. This method is often used to make a direct test of the graduations on an ammeter.

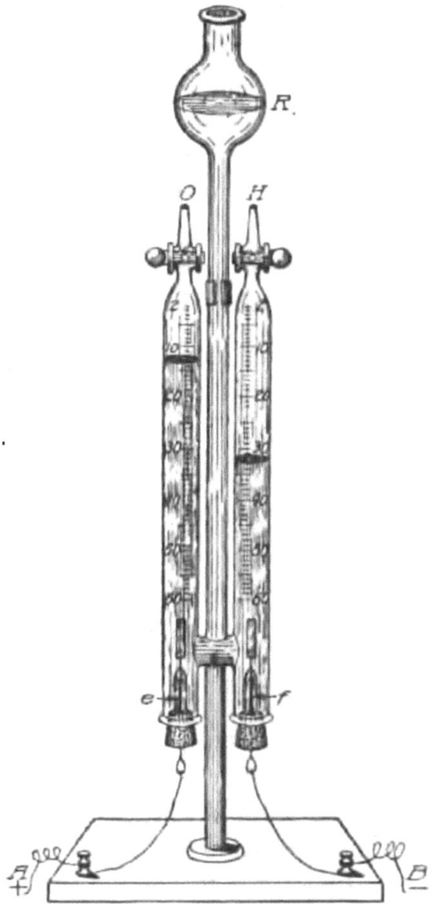

Fig. 91. Measuring Relative Quantity of Hydrogen and Oxygen on Decomposition

Data Illustrating an Ammeter Test

The current supplied by a storage battery was run through an ammeter, a hydrogen voltameter, and a control resistance, and the ammeter reading was kept constant at .330 ampere, or 330 milliamperes, for 10 minutes. The following values were obtained:

$V = 25.4$ cc. $B = 745$ mm.
$t = 20°$ C. $h = 180$ mm.

$$p = 17.4 \text{ mm.}$$

$$P = 745 + \frac{180}{12} - 17.4 = 742.6 \text{ mm.}$$

$$V_0 = 25.4 \times \frac{742.6}{760} \times \frac{273}{293} = 23.12 \text{ cc.}$$

$$\text{coulombs} = \frac{V_0}{.1156} = \frac{23.12}{.1156} = 200$$

$$\text{amperes} = \frac{\text{coulombs}}{\text{seconds}} = \frac{200}{600} = .333$$

or $$\text{milliamperes} = 333$$

Error in ammeter at this part of the scale = −3 milliamperes
Correction to apply to ammeter reading at this part of the scale = +3 milliamperes

82. International Method of Measuring the Ampere. Silver Voltameter. In 1834, Faraday found that a given current of electricity, flowing for a given time, always deposits the same amount of a given element from a solution, whatever be the nature of the solution which contains the element. For example, one ampere always deposits in one hour 4.025 g. of silver, whether the solution is of silver nitrate, silver cyanide, or any other silver compound. This is equivalent to .001118 g. or 1.118 milligrams in one second, so that the International Electrical Congress held at Chicago in 1893 defined *the ampere to be the strength of current which will deposit .001118 g. of silver per second.* The apparatus used to deposit the silver, Fig. 92, is called a *silver voltameter* or *coulombmeter* since it really measures coulombs. The cathode C is made of silver or platinum. The anode A is a disk of silver. The porous cup P similar to that in a Daniell cell separates the anode and cathode and prevents pieces (especially pieces of peroxide) from falling off the anode

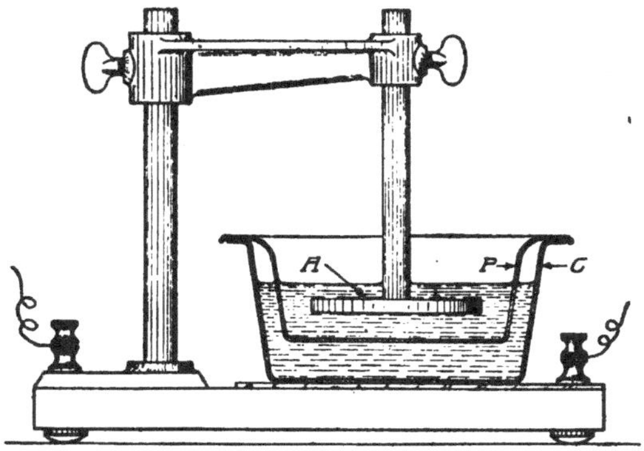

Fig. 92. Section of Silver Voltameter in Action

onto the cathode and thus introducing an error into the true weight of the silver deposited on the cathode. The form of coulombmeter used in the U. S. Bureau of Standards is shown in Fig. 93. After a test, the silver anode carries a crust of disintegrated silver as shown

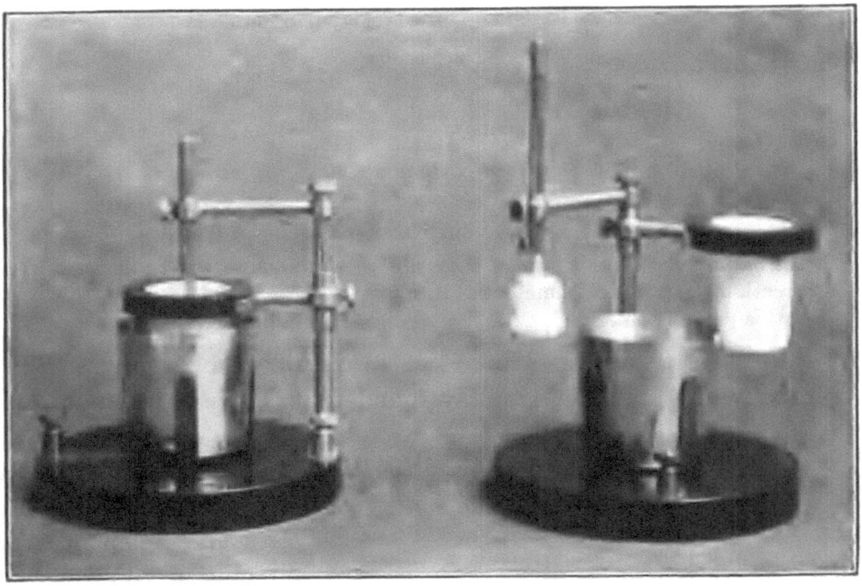

Fig. 93. Type of Silver Voltameter Used in U. S. Bureau of Standards Showing Condition of Anode after a Test

Courtesy of U. S. Bureau of Standards, Washington, D. C.

in the right view of Fig. 93. The pure metallic silver has been deposited on the platinum dish.

Illustrative Example—Ammeter Test Using Silver Voltameter

Mass of platinum cathode C	$=40.1036$ g.
Mass of platinum cathode C after current flowed 30 minutes	$=41.1098$ g.
Mass of silver deposited	$=1.0062$ g.

Since

$$\text{g. of silver deposited} = .001118 \times \text{amperes} \times \text{seconds}$$

$$1.0062 = .001118 \times \text{amperes} \times 1800$$

or

$$\text{amperes} = \frac{1.0062}{.001118 \times 1800} = .5$$

or

500 milliamperes

Ammeter read 502 milliamperes

Error in ammeter at this part of the scale = +2 milliamperes

Correction to apply to ammeter reading at this part of the scale = −2 milliamperes

The electrochemical equivalents for several of the elements, that is, the weight in grams deposited by 1 ampere in 1 second or in other words by 1 coulomb, are found in Table I.

TABLE I

Electrochemical Equivalents

Element	Valency	Equivalent g. per coulomb	Element	Valency	Equivalent g. per coulomb
Aluminum	3	.0000936	Iron	3	.0001929
Antimony	3	.0004153	Lead	2	.0010731
Antimony	5	.0002492	Magnesium	2	.0001260
Bismuth	3	.0007185	Mercury	1	.0020788
Cadmium	2	.0005824	Mercury	2	.0010394
Chromium	3	.0001796	Nickel	2	.0003040
Cobalt	2	.0003055	Oxygen	2	.0000829
Copper	1	.0006588	Platinum	2	.0010104
Copper	2	.0003294	Silver	1	.0011180
Gold	3	.0006812	Tin	2	.0006166
Hydrogen	1	.0000104	Tin	4	.0003083
Iron	2	.0002893	Zinc	2	.0003387

The mass M in grams of any one of the above elements which will be deposited by electrolysis is given by the equation

$$M = eIt$$

in which e is the electrochemical equivalent in grams per ampere second, i.e., in grams per coulomb, I is the current in amperes, and t is the time the current is on in seconds.

EXAMPLES FOR PRACTICE

1. It is desired to make for experimental purposes 500 cc. of dry hydrogen at 22° C. and 75 cm. pressure. How long will it take to make the hydrogen by electrolysis using a current of 1.5 amperes? (Take $P = 75$ as p is here equal to zero.) Ans. 43 min. 54 sec.

2. How many grams of hydrogen will be liberated in one hour by electrolysis using a current of 2 amperes? (Use equation, $M = eIt$; also see Table of Electrochemical Equivalents, p. 92.)

Ans. 0.07488 g.

3. If the hydrogen produced in problem 2 were dried and brought to a pressure of 0° C. and 76 cm., its density would be .00008984. How many cc. would it occupy? Ans. 833.3 cc.

4. How long will it take a current of 2 amperes to deposit 2 grams of silver from a solution of silver nitrate?

Ans. 14 min. 54 sec.

5. How much copper will be deposited per day of 24 hours in a large copper-refining vat if a current of 400 amperes flows through the vat between the plates? This method is used extensively in the production of pure copper. (Use electrochemical equivalent for copper corresponding to valency 2.)

Ans. {11384 g. / 25.1 lb.}

6. In calibrating an ammeter, the current which produces a certain deflection is found to deposit 2.5155 g. of silver in 15 minutes on the cathode of a silver voltameter which is connected in series with the ammeter. What was the strength of the current used and what correction should be applied to the ammeter reading at this part of the scale if the ammeter read 2.55 amperes?

Ans. 2.5 amperes
Correction .05 amperes

LAWS OF RESISTANCE

83. Introduction. The different factors which affect the resistance of a conductor were briefly treated in Part I, but are of such importance that they will be more fully discussed in the following pages.

All substances resist the passage of electricity, but the resistance offered by some is very much greater than that offered by others. Metals have by far the least resistance and of these, silver possesses the least of any. In other words, silver is the best conductor. If the temperature remains the same, the resistance of a conductor is not affected by the current passing through it. A current of ten, twenty, or any number of amperes may pass through a circuit, but its resistance will be unchanged with constant temperature. Resistance is affected by the temperature and also by the degree of hardness. Annealing decreases the resistance of a metal.

84. Conductance. Conductance is the inverse of resistance; that is, if a conductor has a resistance of R ohms, its conductance is equal to $\frac{1}{R}$. Thus, we may think of resistance as the *difficulty* which the current meets or overcomes, and conductance as the *ease* with which it flows. The unit of conductance is the mho (the ohm spelled in reverse order). Thus we see that

$$1 \text{ mho} = \frac{1}{1 \text{ ohm}}$$

Consequently, a wire that has a conductance of 3 mhos has a resistance of $\frac{1}{3}$ ohm and a wire that has a resistance of 5 ohms has a conductance of $\frac{1}{5}$ mho, etc.

85. Resistance Proportional to Length. The resistance of a conductor is directly proportional to its length. Hence, if the length of a conductor is doubled, the resistance is doubled, or if the length is divided, say into three equal parts, then the resistance of each part is one-third the total resistance.

Examples. 1. The resistance of 1283 feet of a certain wire is 6.9 ohms. What is the resistance of 142 feet of the same wire?

As the resistance is directly proportional to the length, we have the proportion

$$\text{required resistance} : 6.9 : : 142 : 1283$$

$$\frac{\text{required resistance}}{6.9} = \frac{142}{1283}$$

$$\text{required resistance} = 6.9 \times \frac{142}{1283}$$

$$= .76 \text{ ohm (approx.)}$$

Ans. .76 ohm (approx.)

2. The resistance of a wire having a length of 521 feet is .11 ohm. What length of the same wire will have a resistance of .18 ohm?

As the resistance is proportional to length, we have the proportion

$$\text{required length} : 521 : : .18 : .11$$

$$\frac{\text{required length}}{521} = \frac{.18}{.11}$$

$$\text{required length} = 521 \times \frac{.18}{.11}$$

$$= 852 \text{ feet (approx.)}$$

Ans. 852 feet

86. Resistance Inversely Proportional to Cross-Section. The resistance of a conductor is inversely proportional to its cross-sectional area. Hence, the greater the cross-section of a wire the less is its resistance. Therefore, if two wires have the same length, but one has a cross-section three times that of the other, the resistance of the former is one-third that of the latter. (In hydrostatics this is analogous to saying that, under a given pressure, a water main having a cross-section of three square feet will carry three times as much water, i.e., will offer one third as much resistance, as one having a cross-section of one square foot.)

Examples. 1. The ratio of the cross-sectional area of one wire to that of another of the same length and material is $\frac{257}{101}$. The resistance of the former is 16.3 ohms. What is the resistance of the latter?

As the resistances are inversely proportional to the cross-sections, the smaller wire has the greater resistance, and we have the proportion

$$\text{required resistance} : 16.3 : : 257 : 101$$

$$\frac{\text{required resistance}}{16.3} = \frac{257}{101}$$

$$\text{required resistance} = 16.3 \times \frac{257}{101}$$

$$= 41.5 \text{ ohms (approx.)}$$

Ans. 41.5 ohms

2. If the resistance of a wire of a certain length having a cross-sectional area of .0083 square inch is 1.7 ohms, what would be its resistance if the area of its cross-section were .092 square inch?

Since increasing the cross-sectional area of a wire decreases its resistance, we have the proportion

$$\text{required resistance} : 1.7 :: .0083 : .092$$

$$\frac{\text{required resistance}}{1.7} = \frac{.0083}{.092}$$

$$\text{required resistance} = 1.7 \times \frac{.0083}{.092}$$

$$= .15 \text{ ohm (approx.)}$$

Ans. .15 ohm

As the area of a circle is proportional to the square of its diameter, it follows that the resistances of round conductors are inversely proportional to the squares of their diameters.

3. The resistance of a certain wire having a diameter of .1 inch is 12.6 ohms. What would be its resistance if the diameter were increased to .32 inch?

The resistances being inversely proportional to the squares of the diameters, we have

$$\text{required resistance} : 12.6 :: .1^2 : .32^2$$

$$\frac{\text{required resistance}}{12.6} = \frac{.1^2}{.32^2}$$

$$\text{required resistance} = 12.6 \times \frac{.1^2}{.32^2}$$

$$= \frac{12.6 \times .01}{.1024}$$

$$= 1.23 \text{ ohms (approx.)}$$

Ans. 1.23 ohms

87. Specific Resistance. The specific resistance of a substance is the resistance of a portion of that substance of unit length and unit cross-section at a standard temperature. The units commonly used are the centimeter or the inch, and the temperature that of melting ice. The specific resistance may therefore be said to be the resistance (usually stated in microhms) of a centimeter cube or of an inch cube at the temperature of melting ice. If the specific resistances of two substances are known, then their relative resistance is given by the ratio of the specific resistances.

88. Conductivity is the reciprocal of specific resistance.

Example. A certain copper wire at the temperature of melting ice has a resistance of 29.7 ohms. Its specific resistance (resistance of 1 centimeter cube in microhms) is 1.594, and that of platinum is 9.032. What would be the resistance of a platinum wire of the same size and length of the copper wire, and at the same temperature?

The resistance would be in direct ratio to the specific resistances, and we have the proportion

$$\text{required resistance} : 29.7 :: 9.032 : 1.594$$

$$\text{required resistance} = 29.7 \times \frac{9.032}{1.594}$$

$$= 168.2 \text{ ohms (approx.)}$$

Ans. 168.2 ohms

89. Calculation of Resistance. From the preceding pages, it is evident that resistance varies directly as the length, inversely as the cross-sectional area, and depends upon the specific resistance of the material. This may be expressed conveniently by the formula

$$R = s\frac{L}{A}$$

in which R is the resistance, L the length of the conductor, A the area of its cross-section, and s the specific resistance of the material.

Example. A telegraph relay is wound with 1,800 feet of wire .010 inch in diameter, and has a resistance of 150 ohms. What will be its resistance if wound with 400 feet of wire .022 inch in diameter?

If the wires were of equal length, we should have the proportion

$$\text{required resistance} : 150 :: (.010)^2 : (.022)^2$$

or

$$\text{required resistance} = 150 \times \frac{(.010)^2}{(.022)^2} = 30.99+ \text{ ohms}$$

For a wire 400 feet long, we have, therefore, by direct proportion

$$\text{required resistance} = \frac{400}{1,800} \times 30.99 = 6.88+$$

Ans. 6.88+ ohms

If a circuit is made up of several different materials joined in series with each other, the resistance of the circuit is equal to the sum of the resistances of its several parts. In calculating the resistance of such a circuit, the resistance of each part should first be calculated, and the sum of these resistances will be the total resistance of the circuit.

In Table II are given the resistances of chemically pure substances at 0° centigrade or 32° Fahrenheit in International ohms. The first column of numbers gives the relative resistances when that of annealed silver is taken as unity. For example, mercury has 62.73 times the resistance of annealed silver. The second and third columns give, respectively, the resistances of a foot of wire .001 inch in diameter, and of a meter of wire 1 millimeter in diameter. The fourth and fifth columns give, respectively, the resistance in microhms of a cubic inch and cubic centimeter, that is, the specific resistances.

TABLE II

Resistance of Chemically Pure Substances at 32° Fahrenheit in International Ohms

Metal	Relative Resistance	Resistance of a wire 1 foot long .001 inch in diameter	Resistance of a wire 1 meter long 1 millimeter in diameter	Resistance in Microhms	
				Cubic Inch	Cubic Centimeter
Silver, annealed	1.000	9.023	.01911	.5904	1.500
Copper, annealed	1.063	9.585	.02028	.6274	1.594
Silver, hard drawn	1.086	9.802	.02074	.6415	1.629
Copper, hard drawn	1.086	9.803	.02075	.6415	1.629
Gold, annealed	1.369	12.35	.02613	.8079	2.052
Gold, hard drawn	1.393	12.56	.02661	.8224	2.088
Aluminum, annealed	1.935	17.48	.03700	1.144	2.904
Zinc, pressed	3.741	33.76	.07143	2.209	5.610
Platinum, annealed	6.022	54.34	.1150	3.555	9.032
Iron, annealed	6.460	58.29	.1234	3.814	9.689
Lead, pressed	13.05	117.7	.2491	7.706	19.58
German silver	13.92	125.5	.2659	8.217	20.87
Platinum-silver alloy (⅓ platinum, ⅔ silver)	16.21	146.3	.3097	9.576	24.32
Mercury	62.73	570.7	1.208	37.05	94.06

It should be noted that the resistances in Table II are for chemically pure substances, and these also at 32° Fahrenheit. A very small portion of foreign matter mixed with a metal greatly increases its resistance. An alloy of two or more metals always has a higher specific resistance than that of any of its constituents. For example, the conductivity of silver mixed with 1.2 per cent in volume of gold, will be 59 when that of pure silver is taken as 100. Annealing reduces the resistance of metals.

The following examples are given to illustrate the use of Table II in connection with the formula on page 97, and to show the application of preceding laws.

Examples. 1. From the specific resistance of annealed aluminum as given in the next to the last column of Table II, calculate the resistance given in the second column of figures for that substance.

The resistance in microhms of a cubic inch of annealed aluminum at 32° F. is 1.144, which is equal to .000001144 ohm. The resistance of a wire 1 foot long and .001 inch in diameter is required. In the formula on page 97 we have $s = .000001144$, $L = 1$ foot $= 12$ inches and

$$A = \frac{\pi d^2}{4} = \frac{3.1416 \times .001^2}{4} = .0000007854 \text{ sq. in.}$$

Substituting these values in the formula

$$R = s\frac{L}{A}$$

we have

$$R = .000001144 \times \frac{12}{.0000007854}$$

$$= 17.48 \text{ ohms} \qquad \text{Ans. } 17.48 \text{ ohms}$$

2. The resistance in microhms of a cubic centimeter of annealed platinum at 0° C. is 9.032. What is the resistance of a wire of the same substance one meter long and one millimeter in diameter at the same temperature?

In the formula for resistance, we have the quantities $s = 9.032$ michroms $= .000009032$ ohm; $L = 1$ meter $= 100$ centimeters; and

$$A = \frac{\pi d^2}{4} = \frac{3.1416 \times .1^2}{4} = .007854 \text{ sq. cm.}$$

the diameter being equal to 1 millimeter $= .1$ cm.

Substituting these values we have

$$R = .000009032 \times \frac{100}{.007854}$$

$$= .1150 \text{ ohm} \qquad \text{Ans. } .115 \text{ ohm}$$

3. From Table II the resistance of 1 foot of pure annealed silver wire .001 inch in diameter at 32° F. is 9.023 ohms. What is the resistance of a mile of wire of the same substance 1 inch in diameter at that temperature?

As the resistance of wires is directly proportional to their length and inversely proportional to the squares of their diameters, the required resistance is found by multiplying the resistance per foot by 5,280 and the product by the inverse squares of the diameters.

$$R = 9.023 \times 5{,}280 \times \left(\frac{.001}{.1}\right)^2$$
$$= 4.76 \text{ ohms (approx.)}$$

Ans. 4.76 ohms

4. A mile and one-half of an annealed wire of pure iron has a resistance of 46.1 ohms. What would be the resistance of hard-drawn wire of pure copper of the same length and diameter, assuming each to be at the temperature of melting ice?

The only factor involved in this example is the relative resistance of the two metals. From Table II, annealed iron has 6.460 and hard-drawn copper 1.086 times the resistance of annealed silver. Hence, the resistance of the copper is to that of the iron as 1.086 is to 6.460, and the required resistance is

$$R = 46.1 \times \frac{1.086}{6.460} = 7.75 \text{ ohms (approx.)}$$

Ans. 7.75 ohms

5. If the resistance of a wire 7,423 feet long is 18.7 ohms, what would be its resistance if its length were reduced to 6,253 feet and its cross-section made one-half again as large?

As resistance is directly proportional to the length, and inversely proportional to the area of the cross-section, the required resistance is

$$R = 18.7 \times \frac{6{,}253}{7{,}423} \times \frac{2}{3} = 10.5 \text{ ohms (approx.)}$$

Ans. 10.5 ohms

From the preceding analysis, it will readily be seen that if we represent the resistance of a wire 1 foot long and .001 inch in diameter by k (the values of k are given in column 2 of Table II), the length of the wire by L, and the diameter of the wire in mils (1 mil = .001 inch) squared, by $C.\ M.$ (circular mils), then the resistance of any wire is given by the equation

$$R = k\frac{L}{C.M.}$$

TABLE III

Temperature Coefficients

Material	Rise in R. of 1 Ohm When Heated: 1° F.	1° C.
Platinoid	.00012	.00022
Platinum silver	.00014	.00025
German silver	.00022	.00040
Platinum	.0019	.0035
Silver	.0021	.0038
Copper, aluminum	.0022	.0040
Iron	.0026	.0046

Example. Find the resistance of a hard-drawn copper wire 500 feet long which has a diameter of .020 inch?

Since the diameter is equal to 20 mils

$$C.M. = (20)^2 = 400$$

From Table II, $k = 9.803$ ohms

$$R = \frac{9.803 \times 500}{400} = 12.25+ \text{ ohms}$$

Ans. 12.25+ ohms

90. Resistance Affected by Heating. The resistance of metals depends upon the temperature, and the resistance is increased by heating. The heating of some substances, among which is carbon, causes a decrease in their resistance. The resistance of the filament of a lighted incandescent lamp is half as great as the resistance of the same filament when cold. All *metals*, however, have their resistance increased by a rise in temperature. The percentage increase in resistance with rise of temperature varies with the different metals, and slightly for the same metal at different temperatures. The increase is practically uniform for most metals throughout a considerable range of temperature. The resistance of copper increases about .4 per cent per degree centigrade, or about .22 per cent per degree Fahrenheit. The percentage increase in resistance for alloys is much less than that for the simple metals. Standard resistance coils are therefore made of alloys, as it is desirable that their resistance should be as nearly constant as possible.

The change in resistance of one ohm per degree rise in temperature for a substance is called the *temperature coefficient* for that substance. Table III gives the temperature coefficients for a few substances.

If the resistance of a conductor at a certain temperature is known, the resistance the conductor will have at a higher temperature may be found by multiplying the temperature coefficient for the substance by the number of degrees of increase and by the resistance at the lower temperature, and by adding to this result the resistance at the lower temperature. The product of the temperature coefficient and the increase in the number of degrees gives the increase in resistance of one ohm through that number of degrees, and multiplying this product by the number of ohms gives the increase in resistance for the conductor. The result obtained is practically correct for moderate ranges of temperature.

The above method of calculating the resistance of conductors at increased temperatures is conveniently expressed by the following formula

$$R_2 = R_1(1 + at)$$

where R_2 is the resistance at the higher temperature, R_1 that at the lower temperature, a the temperature coefficient for the substance, and t the change in the number of degrees.

From the preceding formula it follows that if the resistance at the higher temperature is known, that at the lower temperature will be given by the formula

$$R_1 = \frac{R_2}{1 + at}$$

In calculating resistances at different temperatures, the temperature coefficient based on the Fahrenheit scale should be used if the change in the number of degrees is given in degrees Fahrenheit, and that based on the centigrade scale if the change is given in degrees centigrade.

Examples. 1. The resistance of a coil of German-silver wire at 12° C. is 1,304 ohms. What would be its resistance at a temperature of 60° C.?

From the statement of the example $R_1 = 1304$, $t = 60 - 12 = 48$, and from Table III, $a = .0004$. Substituting these values in the first of the preceding formulas, we have

$$\begin{aligned} R_2 &= 1304(1 + .0004 \times 48) \\ &= 1304 \times 1.0192 \\ &= 1329 \text{ ohms (approx.)} \end{aligned}$$

Ans. 1329 ohms

TABLE IV
American Wire Gage (B. & S.)

No.	Diameter in		Circular Mils	Ohms per 1000 Ft.	No.	Diameter in		Circular Mils	Ohms per 1000 Ft.
	Mils	Mm.				Mils	Mm.		
0000	460.00	11.684	211600.0	.051	19	35.89	.912	1288.0	8.617
000	409.64	10.405	167805.0	.064	20	31.96	.812	1021.5	10.566
00	364.80	9.266	133079.4	.081	21	28.46	.723	810.1	13.323
0	324.95	8.254	105592.5	.102	22	25.35	.644	642.7	16.799
1	289.30	7.348	83694.2	.129	23	22.57	.573	509.5	21.185
2	257.63	6.544	66373.0	.163	24	20.10	.511	404.0	26.713
3	229.42	5.827	52634.0	.205	25	17.90	.455	320.4	33.684
4	204.31	5.189	41742.0	.259	26	15.94	.405	254.0	42.477
5	181.94	4.621	33102.0	.326	27	14.19	.361	201.5	53.563
6	162.02	4.115	26250.5	.411	28	12.64	.321	159.8	67.542
7	144.28	3.665	20816.0	.519	29	11.26	.286	126.7	85.170
8	128.49	3.264	16509.0	.654	30	10.03	.255	100.5	107.391
9	114.43	2.907	13094.0	.824	31	8.93	.227	79.7	135.402
10	101.89	2.588	10381.0	1.040	32	7.95	.202	63.2	170.765
11	90.74	2.305	8234.0	1.311	33	7.08	.180	50.1	215.312
12	80.81	2.053	6529.9	1.653	34	6.30	.160	39.7	271.583
13	71.96	1.828	5178.4	2.084	35	5.61	.143	31.5	342.443
14	64.08	1.628	4106.8	2.628	36	5.00	.127	25.0	431.712
15	57.07	1.450	3256.7	3.314	37	4.45	.113	19.8	544.287
16	50.82	1.291	2582.9	4.179	38	3.96	.101	15.7	686.511
17	45.26	1.150	2048.2	5.269	39	3.53	.090	12.5	865.046
18	40.30	1.024	1624.1	6.645	40	3.14	.080	9.9	1091.865

2. If the resistance of a copper conductor at 95° F. is 48.2 ohms, what would be the resistance of the same conductor at 40° F.? In this case $R_2=48.2$, $t=95-40=55$, and from Table III, $a=.0022$. Substituting these values in the formula on page 102, we have

$$R_1=\frac{48.2}{1+.0022\times 55}=\frac{48.2}{1.121}$$

$$=43 \text{ ohms (approx.)} \qquad \text{Ans. 43 ohms}$$

91. Resistance Tables. Table IV gives the resistance of the most common sizes of copper wire according to the American or Brown and Sharpe (B. & S.) gage. The resistance given is for pure copper wire at a temperature of 75° F. or 24° C.

The first column gives the number of the wire, the second the diameter in thousandths of an inch or mils, and the third the diameter in millimeters. The fourth column gives the equivalent number of wires each one mil or one-thousandth of an inch in diameter. This is called the size of the wire in circular mils and is equal to the square of the diameter in mils. The fifth column gives the ohms per thousand feet, and the resistance per mile is found by multiplying these values by 5.28. Ordinary commercial copper has a conductivity of from 95 per cent to 97 per cent of that of pure copper. The resistance of commercial wire is therefore from 3 per cent to 5 per cent greater than the values given in Table III. The resistance for any metal other than copper may be found by multiplying the resistance given in Table III by the ratio of the specific resistance of the given metal to the specific resistance of copper.

Table V gives the size of the English or Birmingham wire gage. The B. & S. is however much more frequently used in this country. The Brown and Sharpe gage is a little smaller than the Birmingham for corresponding numbers.

TABLE V

Stubs' or Birmingham Wire Gage (B. W. G.)

No.	Diameter in		No.	Diameter in		No.	Diameter in	
	Mils	Mm.		Mils	Mm.		Mils	Mm.
0000	454	11.53	8	165	4.19	18	49	1.24
00	380	9.65	10	134	3.40	20	35	0.89
1	300	7.62	12	109	2.77	24	22	0.55
4	238	6.04	14	83	2.11	30	12	0.31
6	203	5.16	16	65	1.65	36	4	0.10

EXAMPLES FOR PRACTICE

1. What is the resistance of an annealed silver wire 90 feet long and .2 inch in diameter at 32° F? Ans. .02+ ohm

2. What is the resistance of 300 meters of annealed iron wire 4 millimeters in diameter when at a temperature of 0° C.? Ans. 2.31+ ohms

3. What is the resistance of 2 miles of No. 27 (B. & S.) pure copper wire at 75° F.? Ans. 565+ ohms

4. The resistance of a piece of copper wire at 32° F. is 3 ohms. What is its resistance at 49° F.? Ans. 3.11+ ohms

5. The resistance of a copper wire at 52° F. is 7 ohms. What is its resistance at 32° F.? Ans. 6.70+ ohms

6. What is the resistance of 496 ft. of No. 10 (B. & S.) pure copper wire at 45° F.? Ans. .483+ ohm

7. What is the conductance of a wire which has a resistance of .05 ohm? Ans. 20 mhos

8. The resistance of a certain piece of German-silver wire is 180 ohms. What will the resistance be of another piece of the same length but of three times the diameter? Ans. 20 ohms

9. How many feet of No. 40 copper wire would it take to have the same resistance as a mile of a street-car trolley line made of No. 0000 copper wire? Ans. .247 ft., or 2.96 in.

MEASUREMENT OF P.D. AND E.M.F.

92. Measurement of P.D. In Section 37, Part I, the volt was defined as the p.d. between two charged bodies when it requires $\frac{1}{300}$ of an erg of work to carry a unit plus electrostatic charge from one body to the other against the electrical field existing between the two bodies. If we apply the same definition to the case of a circuit in which an electric current is flowing, Fig. 94, and remember that there is a continuous fall in potential along the wire in going from the plus electrode of the battery, or other generator, on the

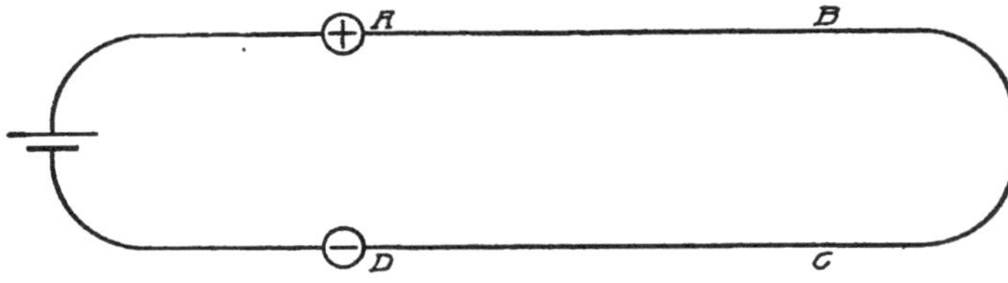

Fig. 94. Simple Battery Circuit Illustrating Fall of Potential

circuit $ABCD$ to the negative electrode, it is evident that a p.d. of 1 volt between two such points as A and D exists if it takes $\frac{1}{300}$ of an erg of work to carry a unit plus electrostatic charge from D to A. This work so done may be proved to be the same whether the unit plus charge is moved directly from D across to A or taken along the path $DCBA$. On the other hand, the work done by a unit plus charge in flowing from A to D along the path $ABCD$ will be $\frac{1}{300}$ of an erg when the potential drop (p.d.) from A to D along the circuit $ABCD$ is 1 volt.

If the current, in going along the path $ABCD$, flows through nothing besides a conductor, then all of the work done by the current will be used in overcoming the friction in the wire (resistance) and will appear as heat energy. However, if a little motor or an electrolytic cell is interposed in the path $ABCD$, then the work done by the electric current will be equal to the mechanical equivalent of the heat produced plus the work done by the motor, or plus the energy required to produce the chemical changes in the electrolytic cell.

93. Absolute Electrometer and Electrostatic Voltmeter. In the diagram shown in Fig. 95, it is evident that, since the charges on the plates *a* and *b* are static, and since these plates are connected to the points *A* and *B*, the potentials of the plates *a* and *b* must be the same as the potentials of the points *A* and *B*. Hence the p.d. between *a* and *b* is the same as the p.d. between *A* and *B*. If the apparatus is balanced with no current flowing in the circuit, and then again when the current is flowing, the weight added to maintain equilibrium when the current is made will give the force

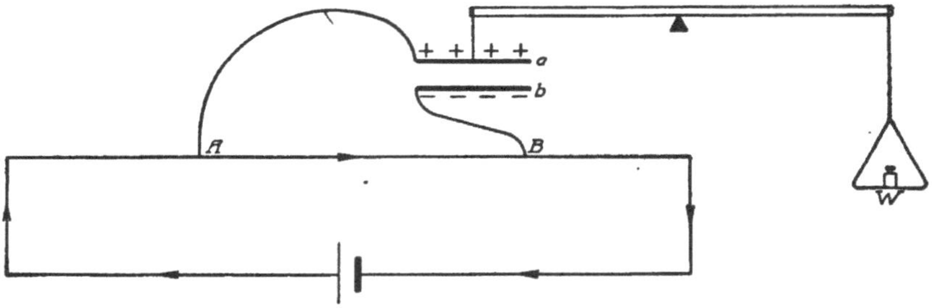

Fig. 95. Diagrammatic Illustration of Principle of Absolute Electrometer

of attraction between the plates *a* and *b*. From this force the strength of the electrostatic field between the plates is easily computed. This strength of field multiplied by the distance between plates in cm. will give the number of ergs of work necessary to carry a unit plus electrostatic charge from *b* to *a*. This number of ergs divided by $\frac{1}{300}$ (or multiplied by 300) will give the p.d. between *a* and *b* and consequently between *A* and *B* in volts.

The above explanation illustrates the principle of the "absolute electrometer" which is used to make direct measurements of p.d. The principle of the electrostatic voltmeter, Fig. 43, Part I, is very similar to the above. Such a voltmeter, however, is calibrated by attaching its terminals to a source whose p.d. is accurately known, and marking the given p.d. opposite the deflection of the movable pointer of voltmeter. By marking several known p.d.'s on the instrument at intervals along the scale, and then dividing the space between these into smaller divisions, the instrument may be very accurately calibrated. Since no current flows through the instrument, it does not tend to lower the p.d. between the points of a circuit to which it is attached as does the ordinary voltmeter, and for this reason, together with the fact that the range of p.d. which it can measure is easily made very large, it is coming more and more into use.

94. Measurement of P.D. by Calorimetric Method. This method consists in measuring the number of calories of heat devel-

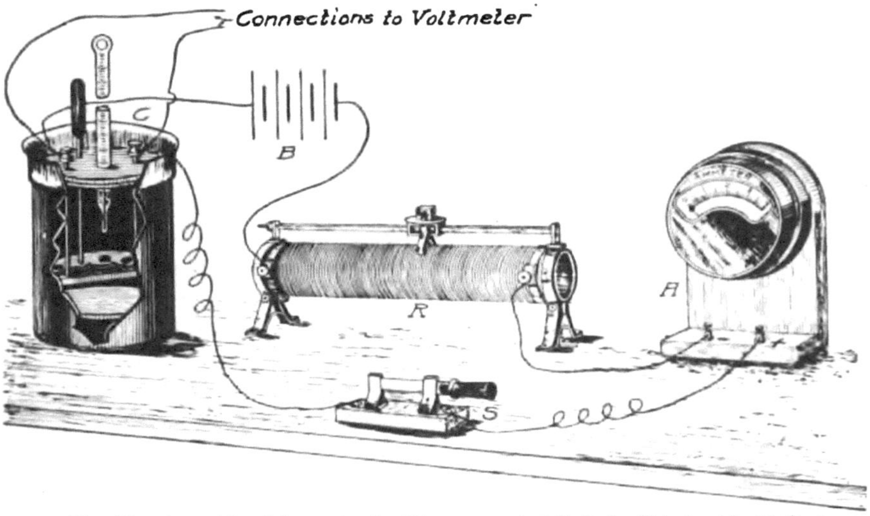

Fig. 96. Assembly of Apparatus for Measurement of P. D. by Calorimetric Method

oped in a coil such as that immersed in the calorimeter *C*, Fig. 96, when a known quantity of electricity is measured in coulombs. (Remember that a coulomb is the quantity of electricity which passes through a conductor in 1 sec. when a current of 1 ampere is flowing. It is also equivalent to 3 billion, i.e., 3×10^9, electrostatic units of electricity passing through the conductor per sec.) If we let the current flowing along the circuit *ABCD*, Fig. 94, be 1 ampere, then in one second, 3 billion electrostatic units of electricity pass from *A* around to *D*. If the p.d. between *A* and *D* is 1 volt, then the work done by each electrostatic unit in flowing from *A* to *D* is $\frac{1}{300}$ erg, see Section 92, and the work done by 3 billion electrostatic units, i.e., by 1 coulomb or 1 ampere-second, will be 3 billion times $\frac{1}{300}$, or 10 million ergs. Thus the work done in 1 second when 1 ampere flows between two points whose p.d. is 1 volt, is 10 million (10^7) ergs, or 1 joule. Evidently when the p.d., current, and time are not unity, the work will be given by the equation

$$\begin{aligned}\text{joules} &= \text{volts} \times \text{amperes} \times \text{seconds}\\ &= \text{p.d.} \times I \times t\end{aligned}$$

In the following test, the circuit $ABCD$ of Fig. 94 is replaced by the platinum coil immersed in the calorimeter C, Fig. 96. The current flowing from the storage battery B is regulated and kept constant by the variable resistance R and is accurately measured by the ammeter A which has been previously tested by comparison with the silver voltameter, Section 82. The voltmeter to be calibrated, or tested if previously calibrated, is shunted across the platinum coil. The mass of the water in grams plus the water equivalent of the thermometer, stirrer, coil, and calorimeter, multiplied by the number of degrees centigrade through which the water is raised, gives the number of calories of heat developed by the current. And since it takes 4.19 joules of work to produce 1 calorie, the mechanical equivalent of this heat will be

$$\text{joules} = 4.19 \times \text{calories} = 4.19\,H$$

and equating the work done by the current in joules to the mechanical equivalent of the heat produced, also in joules, we have

$$\text{volts} \times \text{amperes} \times \text{seconds} = 4.19 \times \text{calories}$$

or

$$\text{p.d.} \times I \times t = 4.19\,H$$

Hence the p.d. as read on the voltmeter, if correct, should be

$$\text{p.d.} = \frac{4.19\,H}{I \times t}$$

Therefore

$$\text{p.d.} = \frac{4.19\ (332 \times 11)}{5.1 \times 100} = 30.0 \text{ volts}$$

The average reading of the voltmeter was 30.1 volts. It therefore read .1 volt, or $\frac{1}{3}$ per cent too high.

The computation of p.d. is based on the following data taken in testing a voltmeter:

Temperature of room	= 18.5° C.
Temperature of water (at make of circuit) ..	= 13.0° C.
Temperature of water (at break of circuit)*.	= 24.0° C.
Weight of water+calorimeter	=425.2 g.
Weight of calorimeter.....................	=105.4 g.
Weight of water...........................	=319.8 g.
Water equivalent of stirrer†...............	= .6 g.
Water equivalent of thermometer†.........	= .9 g.
Water equivalent of coil and electrodes†....	= .8 g.
Water equivalent of calorimeter†..........	= 9.9 g.
Total water equivalent+water............	=332.0 g.
Current..................................	= 5.1 amperes
Time between make and break of circuit ...	=100.0 sec.

MEASUREMENT OF E.M.F. OF A CELL

95. Potentiometer Method. The potentiometer method of comparing the e.m.f. of any cell with that of a standard cell is not

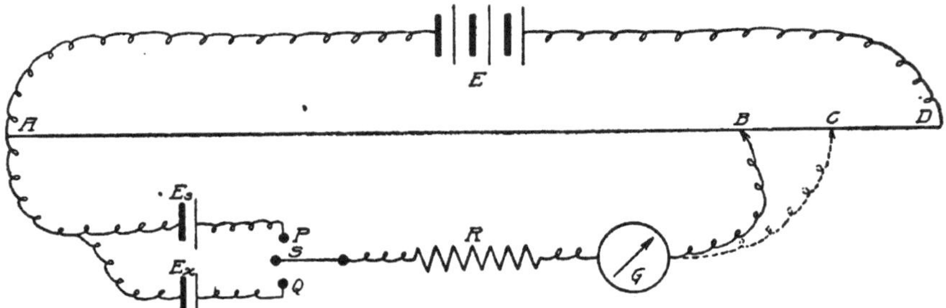

Fig. 97. Diagrammatic Illustration of Potentiometer Method of Comparing E M F. of Cells

only the most convenient but also the standard method employed in finding the e.m.f. of a cell. By keeping the temperature and other conditions of the circuit sufficiently constant, this method can be made to give the e.m.f. of a cell accurately to the fifth decimal place. A storage battery E or any other constant source of e.m.f. which is somewhat higher than that of the cells to be compared is connected to the wire AD, Fig. 97, which should have a resistance sufficient to enable it to carry the current flowing in it without appreciable heating. Evidently the e.m.f. of E must be high enough to

* This temperature should not be recorded after the break of the circuit until the thermometer ceases to rise, the water having been thoroughly stirred during the experiment; and the temperature should be as much above that of the room as that of the water was below that of the room when the circuit was made.

† These are obtained by multiplying the mass of each by its specific heat.

maintain a p.d. between A and D greater than the e.m.f. of either cell. The standard cell E_s, and the cell E_x, whose e.m.f. is unknown, are connected to a double throw switch S, so that either one can be put in series with the resistance R, the galvanometer G, and a movable wire which can be put in contact with the wire AD at any such point as B. Since B is at a higher potential than A, if the switch were on P, a current would flow along $BGRP$ to A were it not for the cell, E_s; if, however, the e.m.f. of E_s is exactly equal to the p.d. from B to A, the electrical pressure tending to send a current in the opposite direction, i.e., along $APRG$ to B, will exactly offset the p.d. from B to A along the potentiometer wire, so that no current will flow in the galvanometer circuit. Evidently under these conditions the e.m.f. of E_s is exactly equal to the p.d. from B to A and is proportional to the length of the wire AB. When the movable wire is to the right of B, the p.d. along the wire will be greater than that of E_s and a current will flow through the galvanometer to the left; likewise, if the movable wire is to the left of B, the e.m.f. of E_s would be greater than the p.d. along BA and a current would flow through the galvanometer to the right. Evidently, then, some current will be drawn from E_s while the position of B for no deflection of the galvanometer is being found. In order then to protect E_s and also to prevent polarization, this current should be made extremely small by having R very large, say 20,000 to 30,000 ohms. This large resistance will also protect the galvanometer G. When the point of balance B has been found, R may be made very small or even cut out of the circuit entirely, thus enabling one to locate the point B more accurately. In exactly the same way, the point C is found when the switch is on Q thus cutting E_s out of the circuit and throwing E_x into the circuit. The e.m.f. of E_x is then equal to the p.d. from C to A, and is proportional to the length of the wire AC. Then, with a perfectly uniform wire, the following equation evidently is true:

$$\frac{E_s}{E_x} = \frac{\text{p.d. from } B \text{ to } A}{\text{p.d. from } C \text{ to } A} = \frac{\text{length } AB}{\text{length } AC}$$

The following data taken to find the e.m.f. of a given Leclanché cell by comparison with a Weston standard cell will serve to illustrate. The temperature of the room was 25° C. The e.m.f. of the Weston standard cell at 20° C. is 1.0183 volts and at any other temperature near that is given by the equation:

$$E=1.0183-0.00004(t-20)$$

Hence $E_s=1.0181$ volts. With E_s in the circuit, the galvanometer showed no deflection when AB was 40.05 cm. When the Leclanché cell was in the circuit, a similar balance was obtained when AC was 61.95 cm. Substituting these values in the above equation gives

$$\frac{1.0181}{\text{e.m.f. of Leclanché cell}}=\frac{40.05}{61.95}$$

$$\text{Hence the e.m.f. of the Leclanché cell}=\frac{1.0181\times 61.95}{40.05}$$

$$=1.575 \text{ volts}$$

Ans. 1.575 volts

EXAMPLES FOR PRACTICE

1. If in Fig. 94, the work required to carry a unit+ electrostatic charge from D to A is 0.01 of an erg, what is the p.d. between A and D along $ABCD$? Ans. 3 volts

2. (a) If a current of 1.5 amperes flows in the circuit of Example 1, how many ergs of work will be done by the current in 5 sec.? (Note.—An ampere equals 3 billion electrostatic units per second passing through the circuit.) Ans. 225,000,000 ergs

(b) To how many joules of work is this equivalent?

Ans. 22.5 joules

(c) What is the rate of expenditure of energy in the portion of the circuit $ABCD$, in watts? (Note.—1 watt equals 1 joule per sec.) Ans. 4.5 watts

(d) In any portion of an electric circuit how may we obtain the rate of expenditure of energy in watts?

Ans. (p.d. in volts)×(I in amperes)

3. With a standard Weston cell at E_s, Fig. 97, the galvanometer showed no deflection when AB was 35.15 cm., the temperature of the room being 22.5° C. When the switch S was thrown over to Q with a dry cell at E_x there was no deflection of the galvanometer when AC was 51.04 cm. Find the e.m.f. of the dry cell used. First find the e.m.f. of the Weston cell at the given temperature.

Ans. { e.m.f. of Weston cell at 22.5° C. = 1.0182 volts
e.m.f. of dry cell used = 1.478 volts

4. The following data was taken on a voltmeter test by the calorimetric method of Section 94. Compute the p.d. across the coil used; compare this p.d. with the voltmeter reading, and give the error of the voltmeter for this part of the scale in volts.

Temperature of room	20° C.
Temperature of water (at make of circuit)	11.8° C.
Temperature of water (at break of circuit)	28.0° C.
Weight of water + calorimeter	481.2 g.
Weight of calorimeter	110.1 g.
Water equivalent of stirrer	1.1 g.
Water equivalent of thermometer	1.2 g.
Water equivalent of coil and electrodes	.9 g.
Water equivalent of calorimeter	10.5 g.
Ammeter reading	7.2 amperes
Average voltmeter reading	40.4 volts
Time between make and break of circuit	90 sec.

Ans. { Computed p.d. across coil is 40.3 volts
Error in voltmeter is +.1 volt

MEASUREMENT OF RESISTANCE

96. Resistance by Substitution. It is evident from Ohm's law, $I=\frac{E}{R}$, or amperes $=\frac{\text{volts}}{\text{ohms}}$, that if the current I and the e.m.f. E of two circuits are the same, then the resistance R of each circuit must be the same. Let the current from the battery E, Fig. 98,

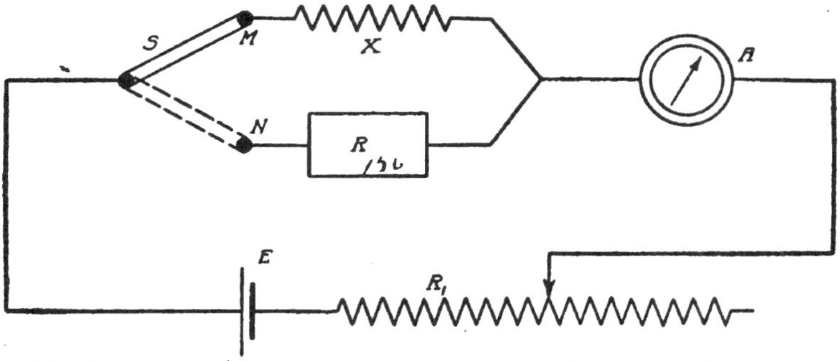

Fig. 98. Diagram of Circuit for Measurement of Resistance by Substitution

be passed through the coil whose resistance X is to be measured through an ammeter A and through a variable control resistance R_1. It is evident that if for a second circuit we use the same circuit, except that we replace X by an adjustable known resistance R, then R will be equal to X when the proper resistance has been taken out of R to just make the ammeter reading the same as it was when X was in the circuit. In taking the data, then, one has only to observe the ammeter reading when the switch S is on M; then to turn the switch on to N and adjust R until the ammeter reads the same as before. Since the cell E in the meantime may have polarized somewhat, the switch should be thrown back to M to see that the current has not changed. If it has, the ammeter reading should again be carefully noted and the switch again thrown over to N and R adjusted to give this latter ammeter reading. These adjustments should be repeated until, when the switch is thrown from M to N,

Fig. 99. Standard Plug-Type Resistance Box and Wheatstone Bridge
Courtesy of Leeds and Northrup Company, Philadelphia

the ammeter reading does not change. Under these conditions, $R = X$. When the circuit is first closed through X, all of the control resistance R_1 should be reduced until a convenient ammeter reading is obtained, using a current which is not large enough to heat X appreciably. The contact points of the switch should be scrupulously clean, as a poor contact has the same effect as putting resistance into the circuit. The points may be cleaned with benzine. The plugs of the resistance box should also be thoroughly cleaned with benzine. One of the most common forms of resistance box which may be used for this work is shown in Fig. 99, the principle of which is illustrated in Fig. 100. The current in passing through the box, from P to Q, or vice versa, must traverse the coil below the plug B, but not the plugs A and C since these plugs are in. Hence, at any time the resistance of the box through which the current flows is given by the sum of the ohms marked on the box by each plug which is out. The low resistances are made of short pieces of heavy manganin wire, the higher ones of longer pieces of fine wire, which, when a considerable length has to be used, are wound on a spool such as *S* in the figure. Manganin wire is used because it has an extremely low temperature coefficient, about .00001 per degree C., and because, with proper use, its resistance will remain practically constant for years. For these reasons, its use has been recommended for all standard resistances by the German *P*hysikalische Technische Reichsanstalt which corresponds to our Bureau of Standards at Washington.

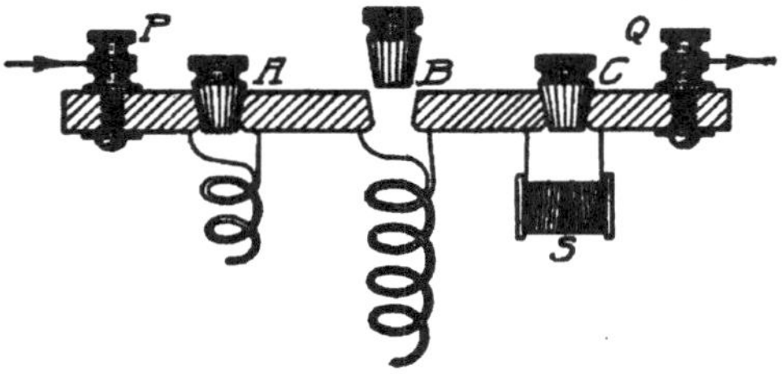

Fig. 100. Section Showing Typical Construction of Resistance Box

97. Resistance by Voltmeter-Ammeter Method. Whenever an accurate voltmeter and ammeter are available, the following method for finding the value of an unknown resistance will be found very convenient and expeditious, and is the method much used by the practical engineer. If the accuracy of the instruments is not known, they should be compared with standard instruments and their errors noted, or else should be calibrated according to one of the methods previously given. The conductor whose resistance *X* is to be found is connected in series with a battery *E*, or other generator, and an ammeter *A*, and the voltmeter is shunted across *X* as in Fig. 101. According to Ohm's law, $I=\frac{\text{p.d.}}{R}$, we then have $R=\frac{\text{p.d.}}{I}$, or $\text{ohms}=\frac{\text{volts}}{\text{amperes}}$; so that the resistance of the coil *X* in

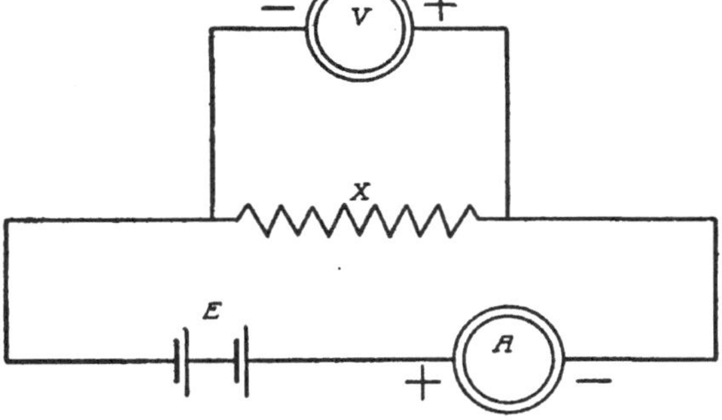

Fig. 101. Diagram for Obtaining Resistance by Voltmeter-Ammeter Method

ohms is simply the voltmeter reading in volts divided by the ammeter reading in amperes. Obviously the ammeter, when connected as above, registers the combined current flowing through the conductor X and the voltmeter V, and we have applied Ohm's law as though all of this current flowed through X and none through the voltmeter; however, the error thus introduced is negligible, providing the current carried by the *voltmeter* is very small in comparison to that carried by X, and this will be the case when X is small in comparison to the voltmeter resistance. If X is more than 1 per cent of the voltmeter resistance and results are to be expected within 1 per cent, the apparatus must then be connected as in Fig. 102. We are now measuring a resistance X which is very large in comparison to the resistance of the *ammeter* so that the p.d. across X is practically the same as it is across both X and A, and the ammeter registers the current flowing through X only. Hence we may apply Ohm's law as before:

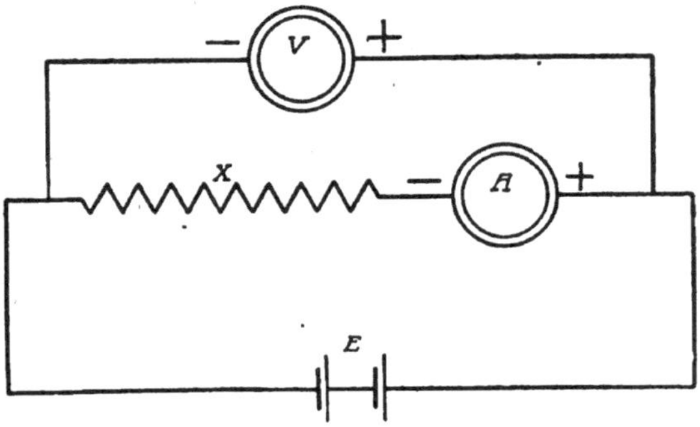

Fig. 102. Diagram of Circuit for Method of Fig. 101 When Unknown Resistance Approaches That of Voltmeter

$$X\text{ (in ohms)} = \frac{\text{p.d. (in volts)}}{I\text{ (in amperes)}}$$

The value of X thus obtained evidently includes the resistance of the ammeter, which, however, may usually be neglected in any practical problem of this sort. However, if one wishes to correct X by subtracting the ammeter resistance, it also may be found by the same method. Connect the voltmeter across the ammeter as in Fig. 103. Vary the resistance of R until the ammeter reading is

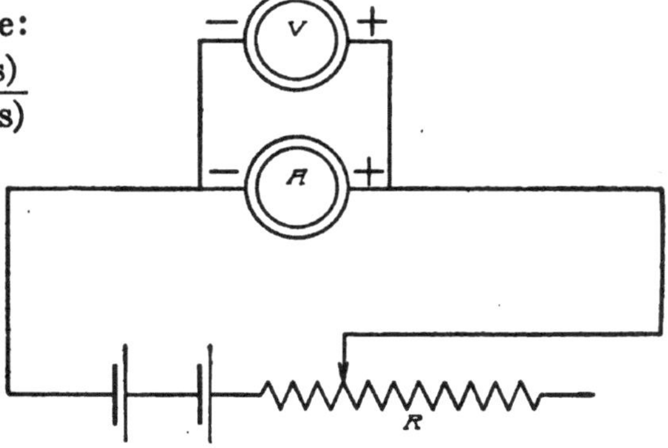

Fig. 103. Diagram of Circuit for Finding the Resistance of Ammeter

as large, or nearly as large, as the ammeter registers. Since the ammeter registers the current through itself and the voltmeter gives the p.d. across the ammeter, evidently the resistance of the ammeter is the voltmeter reading in volts divided by the ammeter reading in amperes. However, the p.d. across the ammeter will be very small since its resistance is small and consequently the voltmeter to be used should be a millivoltmeter, i.e., one reading to thousandths of a volt.

98. Resistance by Wheatstone Bridge. *Discussion of Method.* When a high degree of accuracy in resistance measurements is desired. the Wheatstone bridge is the instrument which is much employed, It is used in testing laboratories when the previous methods are not sufficiently accurate, and when the larger currents and voltages used in these methods are objectionable on account of heating

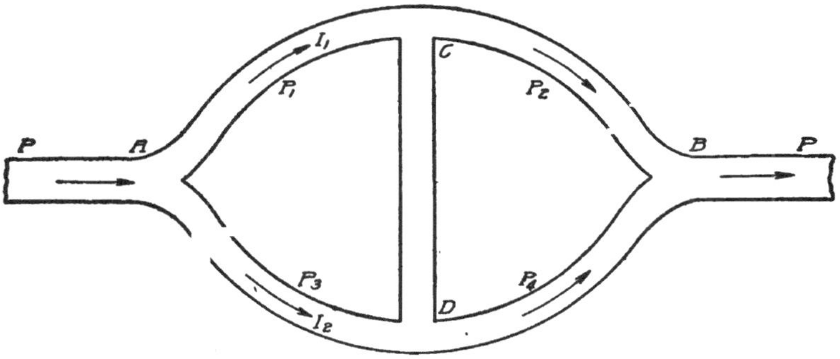

Fig. 104. Water Analogy Illustrating Principle of Wheatstone Bridge

effects and consequent changes in resistance. Its value will be even more appreciated after we have learned how it is used in locating faults and grounds in telegraph and telephone circuits.

The water analogy of Fig. 104 will help to make clear the theory of the Wheatstone bridge. Suppose a current of water which is flowing in the pipe P, divides at A, part of the current I, flowing through the upper branch which consists of pipes P_1 and P_2 to B and on; the other part of the current I_2 flowing through the lower branch which consists of pipes P_3 and P_4 to B and on. The difference in pressure at the ends of the branch ACB is evidently the same as the difference in pressure at the ends of the branch ADB, since their ends have a common meeting point. Since then the decrease in hydrostatic pressure is the same along ACB as along ADB, it follows that for any point C in the upper branch which has a given pressure there must be some point D in the lower branch having the same pressure. Hence, if a pipe be connected between two such points as C and D, no current will flow through it.

Similarly, if we consider a current of electricity flowing from the battery E of Fig. 105, dividing at A, and let I_1 be the part of it which flows through R_1 and R_2 to B and back to E, and let I_2 be the other part which flows through R_3 and R_4 to B and back to E, it is evident that no current will flow from C to D through the galvanometer G, if C and D are at the same electrical pressure,

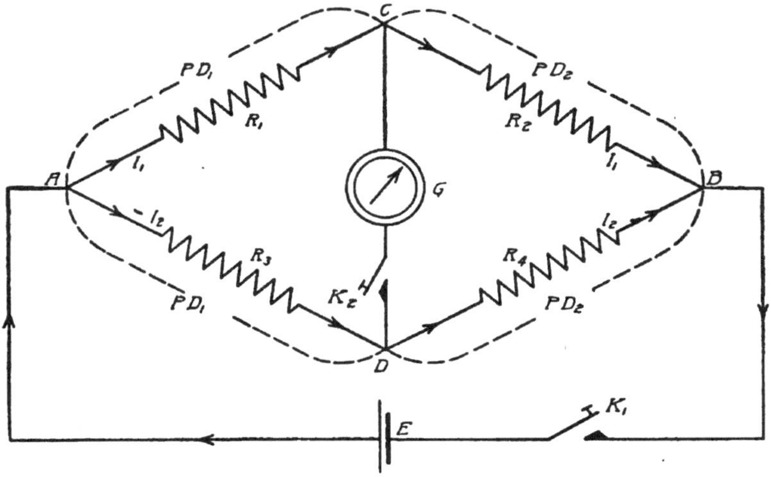

Fig. 105. Standard Diamond Diagram Illustrating Principle of Wheatstone Bridge

i.e., the same potential. Let C and D in Fig. 105, be two such points. Then from Ohm's law we have

$$\text{p.d.}_1 = I_1 R_1 = I_2 R_3$$

and

$$\text{p.d.}_2 = I_1 R_2 = I_2 R_4$$

and dividing the former of these equations by the latter we have

$$\frac{R_1}{R_2} = \frac{R_3}{R_4}$$

If any one of these resistances is unknown, it may be computed if the other three resistances are known. Thus if R_2, R_3, and R_4 are 6, 8, and 12 ohms, respectively, then

$$\frac{R_1}{6} = \frac{8}{12}$$

$$R_1 = \frac{6 \times 8}{12} = 4 \text{ ohms}$$ Ans. 4 ohms

If R_1, R_2, and R_3 are 5, 10, and 15 ohms, respectively, then

$$\frac{5}{10}=\frac{15}{R_4}$$

$$R_4=\frac{15\times10}{5}=30 \text{ ohms} \qquad \text{Ans. 30 ohms}$$

"Slide-Wire Bridge". If we examine carefully the slide-wire bridge shown in Fig. 106 and compare the relative position of resistances, galvanometer, etc., with the corresponding positions which these occupy in Fig. 105, we shall find everything essentially the same. Heavy strips, having negligible resistance, and to which the connections are made, are mounted on a board for the sake of con-

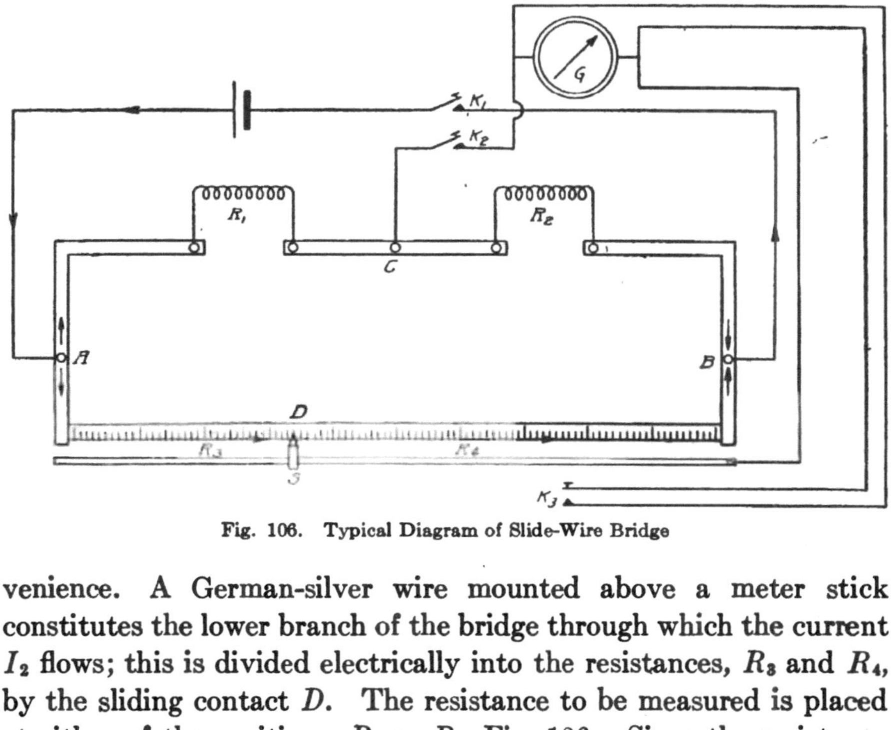

Fig. 106. Typical Diagram of Slide-Wire Bridge

venience. A German-silver wire mounted above a meter stick constitutes the lower branch of the bridge through which the current I_2 flows; this is divided electrically into the resistances, R_3 and R_4, by the sliding contact D. The resistance to be measured is placed at either of the positions, R_1 or R_2, Fig. 106. Since the resistance of a uniform wire is proportional to its length, in place of the ratio $\frac{R_3}{R_4}$, we may substitute in the equation developed for the bridge, the ratio $\frac{\text{length of } R_3}{\text{length of } R_4}$, as read directly on the meter stick beneath the wire. In using the bridge, K_1 should always be closed first, and then K_2 when the bridge is being balanced. The reason for this will be explained when the subject of self-induction is treated.

Fig. 107. Standard Double-Circuit Key
Courtesy of Leeds and Northrup Co., Philadelphia, Pa.

For this purpose a double contact key like that shown in Fig. 107 is very convenient. A disk of insulating material at M, Fig. 108, prevents any connection between the battery and the galvanometer circuits. While balancing the bridge, much time will be saved with most types of galvanometers if a so-called damping key, K_3, is connected to wires running to the two terminals of the galvanometer. When this is closed, the galvanometer will be brought to rest in a comparatively short time. For this purpose a single-contact key is used.

Illustrative Example. The conductor whose resistance X was to be measured was placed at R_1, Fig. 106, and the sliding contact S was placed so that D was on the 50 cm. mark of the meter stick. With no plugs out of the resistance box placed at R_2, the galvanometer was deflected strongly in one direction when K_1 and K_2 were closed. These keys were then opened and the galvanometer brought to rest by closing K_3. Then 10 ohms were taken out of the box at R_2 and a test for a balance showed a deflection of the galvanometer in the same direction. With 20 ohms out of the box, the deflection was still in the same direction, but with 30 ohms it was in the opposite direction. Hence the resistance X must lie

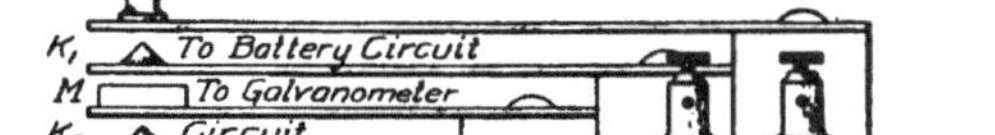

Fig. 108. Double Contact Key Showing Method of Dividing Circuits

between 20 and 30 ohms since R_3 and R_4 are now equal. By repeating the above tests, it was found that 25 ohms at R_2 produced a deflection one way, while 26 ohms produced a deflection the opposite way. Hence the resistance X must lie between 25 and 26 ohms. Then with 25 ohms at R_2, a balance would be obtained by sliding S a little to the right, since R_2 is now less than X and therefore R_4 must be less than R_3. A few trials showed the position of D for a balance to be at 50.8 cm., or the length of R_3 was 50.8 cm., and the length of R_4 was 49.2 cm. Therefore

$$\frac{X}{25}=\frac{50.8}{49.2}$$

$$X=\frac{25\times 50.8}{49.2}=25.81 \text{ ohms}$$

In order to eliminate any bridge errors due to nonuniformity in the wire, etc., the unknown resistance was then placed at R_2 and

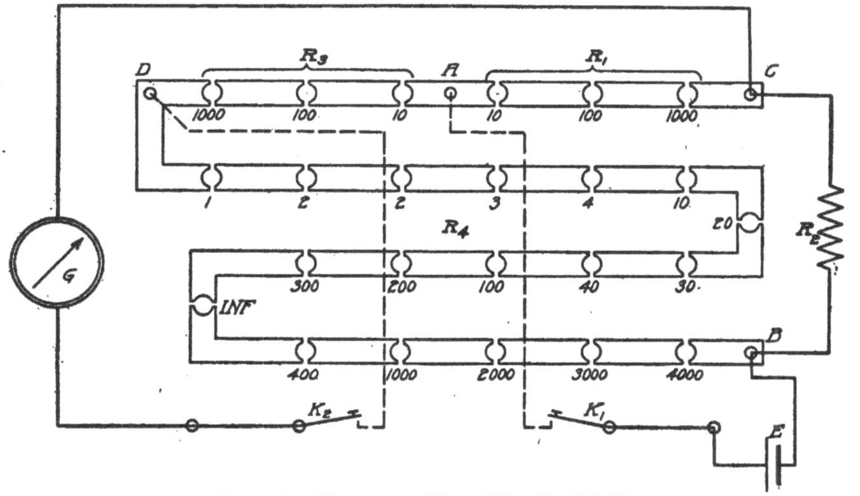

Fig. 109. Diagram of "Post-Office Box" Bridge

the resistance box at R_1 with 25 ohms out. A balance was obtained when D was at 49.12 cm., so that the length of R_3 was 49.12 cm. and that of R_4 was 50.88 cm. Hence, for this balance X is given by

$$\frac{25}{X}=\frac{49.12}{50.88}$$

Where

$$X=\frac{25\times 50.88}{49.12}=25.89 \text{ ohms}$$

The mean of these two determinations gives the correct value of X as 25.85 ohms. Ans. 25.85 ohms

"Post-Office Box" Bridge. The following convenient and much used form of the Wheatstone bridge derived its name from the fact that it was first used by the telegraph department of the British post-office. As in the previously discussed forms of the bridge, it will be noted in Fig. 109, which shows the post-office box diagrammatically, that the current from E flows to A when key K_1 is closed and there divides, part flowing through R_1 and R_2 back to the battery and part through R_3 and R_4 back to the battery. As before, the galvanometer is connected between C and D. Fig. 99 shows the post-office box as it appears in actual form. The resistance to be measured is always placed at R_2. For convenience we will use the equation

$$\frac{R_1}{R_2}=\frac{R_3}{R_4}$$

in the following equivalent form

$$\frac{R_2}{R_4}=\frac{R_1}{R_3}$$

The easiest way to find the unknown R_2, is first to make R_1 equal to R_3, say 10 ohms each, and then adjust R_4 until an increase of 1 ohm will cause an opposite deflection of the galvanometer. We know, then, that the value of the unknown is within 1 ohm of the value of R_4. If, for example, we take as the unknown resistance the coil used in the illustrative example, page 119, when R_4 was 25 ohms, the galvanometer would deflect one way and when R_4 was 26 ohms, it would deflect the opposite way. If, then, R_1 is made 10 ohms and R_3 is made 100 ohms, or a ratio of 10 : 100 is used, it is evident that a balance would now be obtained for some value of R_4 between 250 and 260 ohms. Suppose that 258 ohms at R_4 caused the galvanometer to deflect one way, and 259 ohms caused it to deflect the other way. If, then, we use 10 : 1,000 for the ratio $R_1 : R_3$, we should get a balance when R_4 is between 2,580 ohms and 2,590 ohms. Practically no deflection of the galvanometer was observed when R_4 was 2,586 ohms. Hence the resistance R_2 of the coil is given by

$$\frac{R_2}{2{,}586}=\frac{10}{1{,}000}$$

$$R_2=25.86 \text{ ohms}$$ Ans. 25.86 ohms

EXAMPLES FOR PRACTICE

1. In measuring the resistance of the coils of wire on the electromagnet of an electric bell, "by substitution" it was found that when the bell was placed at *X*, Fig. 98, and the clapper held so that a continuous current flowed through the bell, that the ammeter reading remained the same, when the switch *S* was thrown from *M* to *N*, if *R* was made 1.25 ohms. What was the resistance of the electric bell? Ans. 1.25 ohms

2. An ammeter placed in series with a 16 c-p. carbon filament lamp reads 0.5 ampere while a voltmeter connected across the lamp reads 110 volts. What is the resistance of the lamp? Ans. 220 ohms

3. When the ammeter was placed in series with a 16 c-p. tungsten filament, it read 0.182 ampere, while the voltmeter connected across the lamp read 110 volts. What was the resistance of this lamp? Ans. 604+ ohms

4. An ammeter and a millivoltmeter when connected as in Fig. 103 are found to read 5 amperes and 250 millivolts, respectively. What is the resistance of the ammeter? (1 millivolt = .001 volt.) Ans. 0.05 ohm

5. A voltmeter and a milliammeter, when connected in series to the lighting circuit, read 110 volts and 55 milliamperes, respectively. What is the resistance of the voltmeter? (Remember that when connected this way the ammeter reads the current through the voltmeter and the voltmeter, of course, always reads the p.d. between its own terminals. Also 1 milliampere = .001 ampere.) Ans. Resistance of voltmeter = 2000 ohms

6. In Fig. 102 a telegraph relay, Section 107, is placed at *X*, 2 storage cells at *E*, a milliammeter at *A*, and an ordinary voltmeter at *V*. The ammeter reads 21 milliamperes and the voltmeter reads 4.2 volts. What is the resistance of the windings on the relay? Ans. 200 ohms

7. In Example 6 we neglected the ammeter resistance, why is this permissible? Why, in the same problem, must the voltmeter be shunted across both the relay and the ammeter?

8. Diagram and explain how to find an unknown resistance (a) by substitution; (b) by the voltmeter-ammeter method when the unknown resistance is small in comparison to that of the voltmeter, and also when it is large; (c) by the Wheatstone bridge method.

ELECTROMAGNETISM

99. Magnetic Properties of a Loop. We have seen in Part I, Section 48, that an electrical current is surrounded by a magnetic field the direction of which is given by the right-hand rule—namely, that if the wire is grasped with the right hand, the thumb pointing along the wire in the direction in which the current is flowing, then the fingers will encircle the wire in the same direction as the magnetic lines of force. We have seen also that a loop or a coil of wire through which a current flows produces a magnetic field of the shape shown in Fig. 110. Now if such a loop is suspended* in the manner shown in Fig. 111, while a current from a dry cell *B* is sent through it, the loop will turn and come to rest in an east-and-west plane, so that the face of the loop from which the magnetic lines emerge is toward the north. (See right-hand rule mentioned above and also Fig. 110.) In other words, the loop will be found to behave with respect to the earth or to any other magnet, precisely as though it were a flat magnetic disk whose boundary is the wire, the face which turns toward the north, that is, that from which the magnetic lines emerge, being an *N* pole and the other an *S* pole.

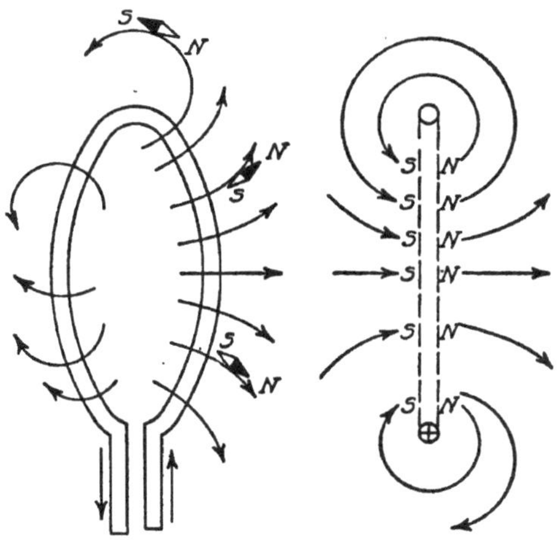

Fig. 110. Diagram of Lines of Force Passing through a Coil Bearing a Current

100. Magnetic Properties of a Helix. If a wire carrying a current be wound in the form of a helix and held near a suspended magnet as in Fig. 112, the coil will be found to act in every respect like a magnet, with an *N* pole at one end and an *S* pole at the other.

This result might have been predicted from the fact that a single

*To make the apparatus of Fig. 111, slide a piece of rubber tubing *R* over a U-shaped tube *U* and then slide the same end of the U-tube into a thistle tube *T* of such a size that the piece of rubber tubing forms a joint between the two tubes, preventing the mercury from running out. The top end of the U-tube should be placed in a flame and allowed to contract, but should still permit the wire dipping into it to fit loosely—this prevents the wire from sticking to the sides of the tube and hence allows it to turn very freely. For the suspension of the loop, untwist a foot of silk twist and use one of the strands.

loop is equivalent to a flat-disk magnet. For when a series of such disks is placed side by side, as in the helix, the result must be the same as placing a series of disk magnets in a row, the N pole of one being directly in contact with the S pole of the next, etc. These poles therefore all neutralize each other except at the two ends. We therefore get a magnetic field of the shape shown in Fig. 113. Such a helix may be made by threading a copper wire through a cardboard or a sheet of celluloid. If iron filings are then sifted onto the cardboard, and the board is tapped gently while a strong current is sent through the helix, the iron filings will become little induced magnets and point in the direction of the magnetic field. It will be noticed that not all of the "magnetic flux" of lines of force is through the coil, but that some of the lines encircle each wire where it passes to the outside of the coil and hence do not add to the magnetic strength of the helix. This flux of lines between the wires is known as the "leakage flux".

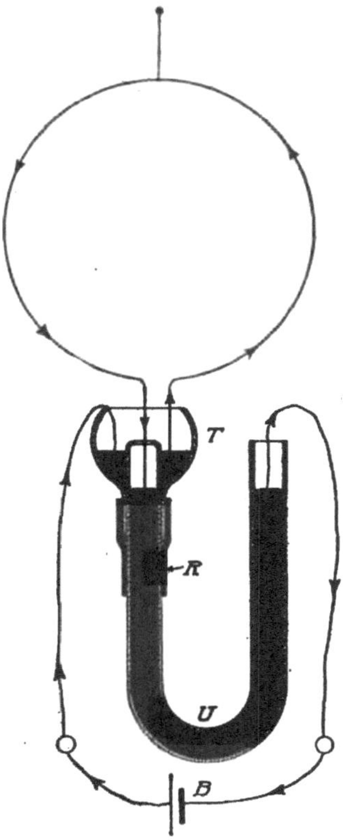

Fig. 111. Section of Apparatus Showing Magnetic Properties of a Suspended Coil

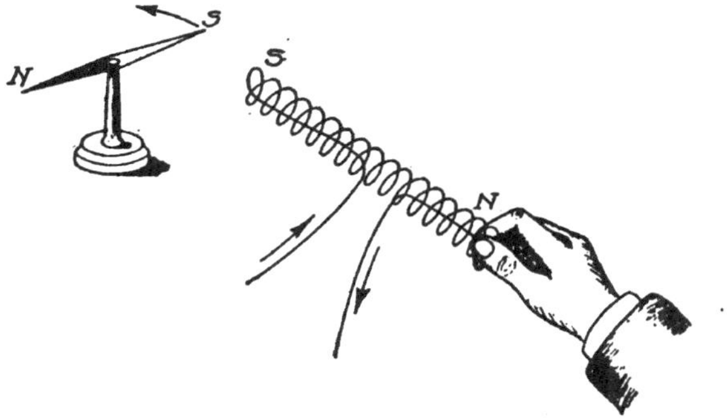

Fig. 112. Experiment Showing Magnetic Properties of Helix

101. Rules for North and South Poles of a Helix. The right-hand rule, as given in Part I, Section 48, is sufficient in every case to determine which is the N and which the S pole of a helix, i.e., from which end the lines of magnetic force emerge from the helix and at which end they enter it. But it is found convenient, in the consideration of coils, to restate the right-hand rule in a slightly different way, thus:

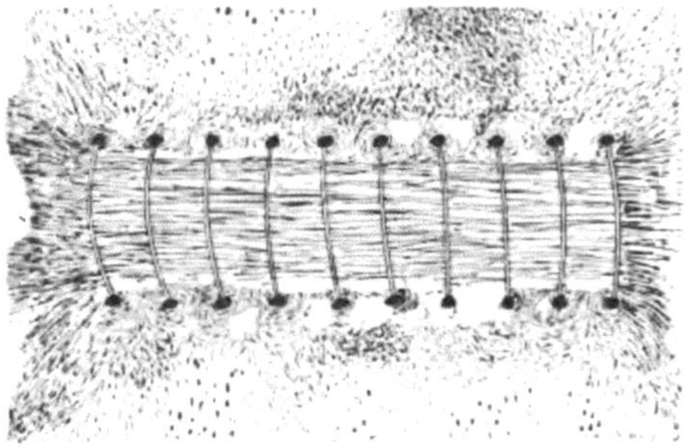

Fig. 113. Magnetic Field of a Helix Bearing a Current

If the coil is grasped in the right hand in such a way that the fingers point in the direction in which the current is flowing in the wires, the thumb will point in the direction of the north pole of the helix, Fig. 114.

Similarly, if the sign of the poles is known, but the direction of the current unknown, the latter may be determined as follows:

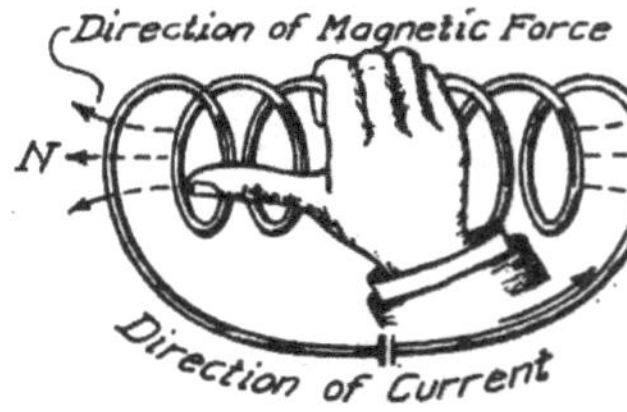

Fig. 114. Diagram Showing Rule for Determining Polarity of a Helix

If the right hand is placed with the palm against the coil and with the thumb pointing in the direction of the lines of force (i.e., toward the north pole of the helix), the fingers will pass around the coil in the direction in which the current is flowing.

102. The Electromagnet. If a core of soft iron be inserted in the helix, Fig. 115, the poles will be found to be enormously stronger than before. This is because the core is magnetized by induction from the field of the helix in precisely the same way in which it would be magnetized by induction if placed in the field of a permanent magnet. The new field strength about the coil is now the sum of the field due to the core and that due to the coil. The introduction of the iron core into the helix almost entirely prevents the "leakage flux" so that this also adds to the strength of the magnet. If the current is broken, the core will at once lose the

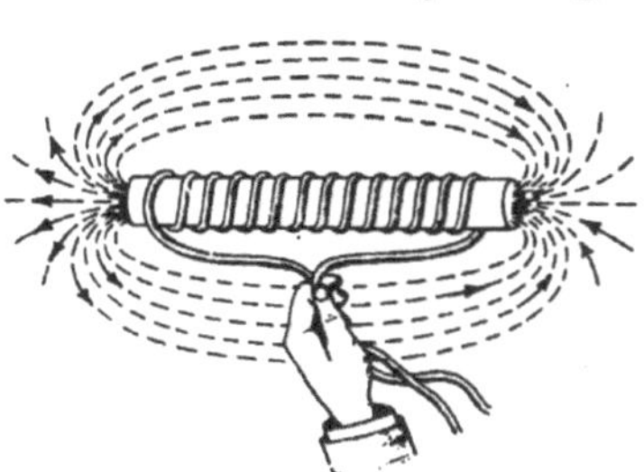

Fig. 115. Simple Electromagnet

greater part of its magnetism. If the current is reversed, the polarity of the core will be reversed. Such a coil with a soft-iron core is called an *electromagnet.*

The strength of an electromagnet can be very greatly increased by giving it such form that the magnetic lines can remain in iron throughout their entire length instead of emerging into air, as they do in Fig. 115. For this reason electromagnets are usually built in the horseshoe form and provided with an armature A, Fig. 116, through which a complete iron path for the lines of force is established, as shown in Fig. 117. The strength of such a magnet depends chiefly upon the cross-section of the iron core and the number of *ampere-turns* which encircle it, the expression *ampere-turns* denoting the product of the number of turns of wire about the magnet by the

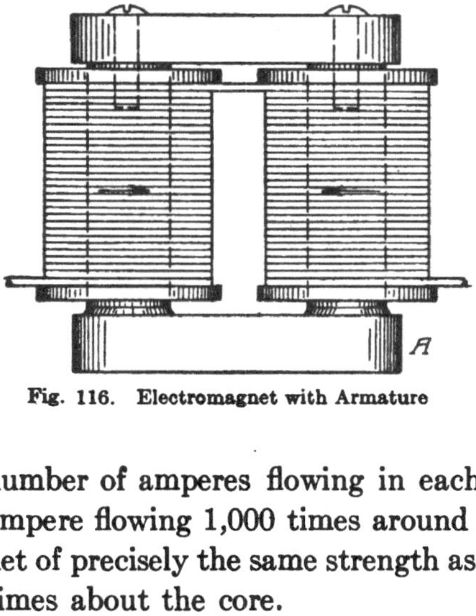

Fig. 116. Electromagnet with Armature

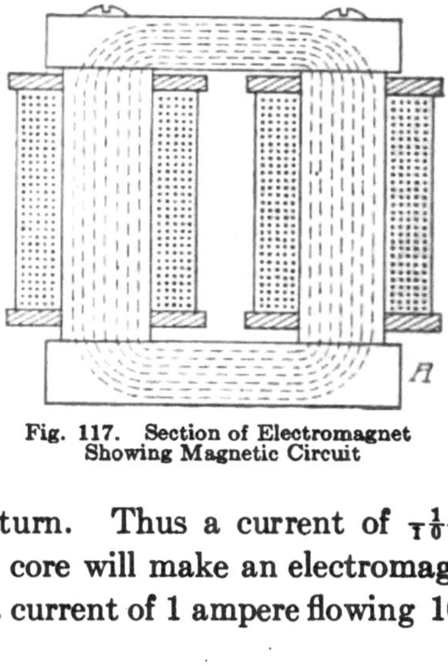

Fig. 117. Section of Electromagnet Showing Magnetic Circuit

number of amperes flowing in each turn. Thus a current of $\frac{1}{100}$ ampere flowing 1,000 times around a core will make an electromagnet of precisely the same strength as a current of 1 ampere flowing 10 times about the core.

APPLICATIONS OF ELECTROMAGNETS

103. The Electric Bell. The electric bell, Fig. 118, is one of the simplest applications of the electromagnet. When the button P is pressed, the electric circuit of the battery is closed, and a current flows in at A, through the magnet, over the closed contact C, and out again at B. But no sooner is this current established than the electromagnet E pulls over the armature a, and in so doing breaks the contact at C. This stops the current and demagnetizes the magnet

E. The armature is then thrown back against *C* by the elasticity of the spring *s* which supports it. No sooner is the contact made at *C* than the current again begins to flow and the former operation is repeated. Thus the circuit is automatically made and broken at *C* and the hammer *H* is in consequence set into rapid vibration against the rim of the bell.

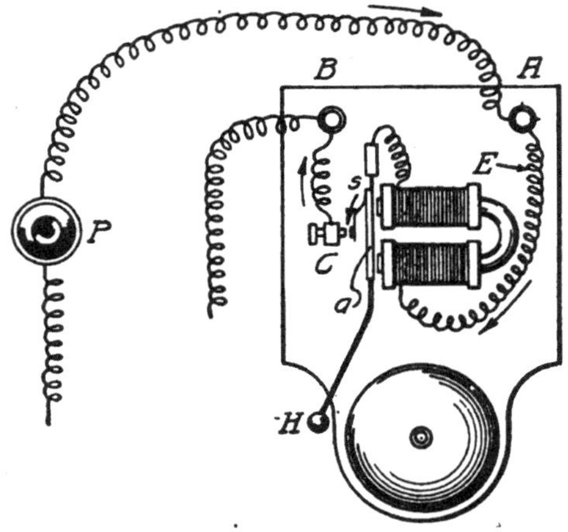

Fig. 118. Adaptation of Electromagnet to Simple Electric Bell

104. Electric Tuning-Fork. It is often desirable to keep a tuning-fork in a uniform state of vibration. For this purpose an electrically driven fork like that of Fig. 119 is used. When the battery is connected to *A* and *B*, the current flows through the electromagnet *M*, thus attracting both prongs of the fork and at the same time breaking the circuit at *C*. The elasticity of the prongs then throws them back making the circuit at *C* and the operation is repeated the number of times per second equal to the frequency of the fork. Thus the operation of such a fork is exactly the same as that of the electric bell.

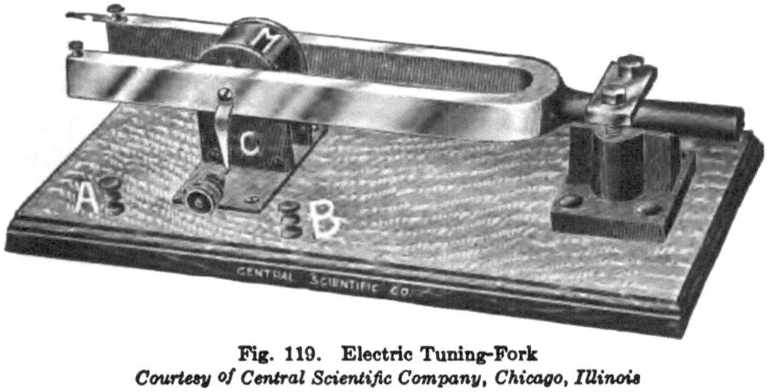

Fig. 119. Electric Tuning-Fork
Courtesy of Central Scientific Company, Chicago, Illinois

105. The Telegraph. The electric telegraph is another simple application of the electromagnet. The principle is illustrated in Fig. 120. As soon as the key *K* at Chicago, for example, is closed, the current flows over the line to New York, we will say. There it passes through the electromagnet *m*, and thence back to Chicago through the earth. The armature *b* is held down by the electromagnet *m* as long as the key *K* is kept closed. As soon as the circuit is broken at *K*, the armature is pulled up by the spring *d*. By means of a clockwork device, the tape *e* is drawn along at a uniform rate beneath the pencil or the pen carried by the armature *b*. A very short time of closing of *K* produces a dot upon the tape, a longer, time a dash. As the Morse, or telegraphic, alphabet consists

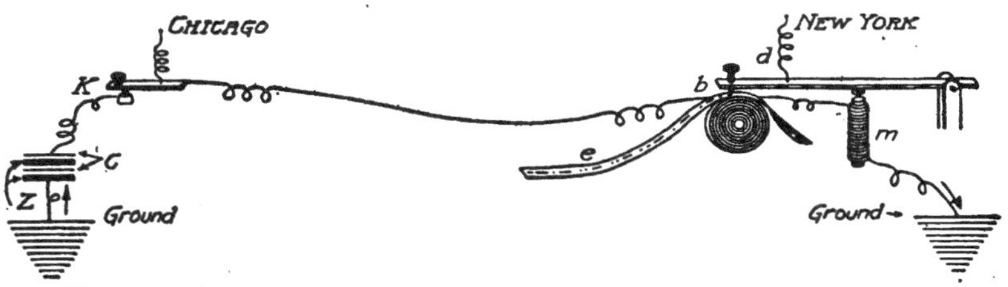

Fig. 120. Simple Diagram of Telegraph Circuit between Chicago and New York

of certain combinations of dots and dashes, any desired message may be sent from Chicago and recorded in New York.

AMERICAN MORSE CODE

A • —	J — • — •	S • • •	2 • • — • •
B — • • •	K — • —	T —	3 • • • — •
C • • •	L ——	U • • —	4 • • • • —
D — • •	M — —	V • • • —	5 — — —
E •	N — •	W • — —	6 • • • • • •
F • — •	O • •	X • — • •	7 — — • •
G — — •	P • • • • •	Y • • • •	8 — • • • •
H • • • •	Q • • — •	Z • • • •	9 — • • —
I • •	R • • •	1 • — — •	0 ———

In modern practice the message is not ordinarily recorded on a tape, for operators have learned to read messages by ear, a very short interval between two clicks being interpreted as a dot, a longer interval as a dash.

The first commercial telegraph line was built between Baltimore and Washington by S. F. B. Morse. It was opened on May 24, 1844, with the now famous message: "What hath God wrought?"

106. The Relay and the Sounder. On account of the great resistance of long lines, the current which passes through the electromagnet is so weak that the armature of this magnet must be made

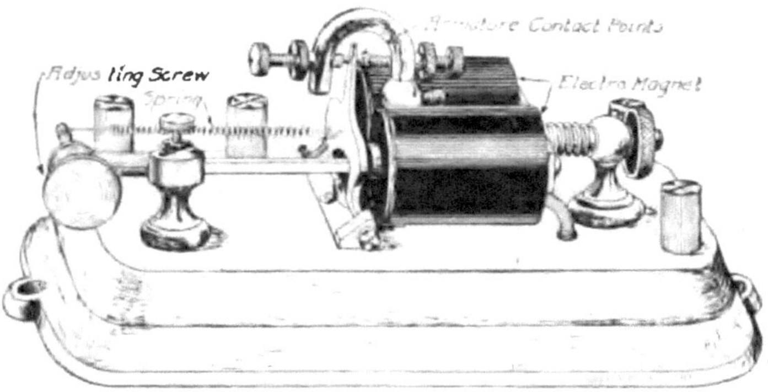

Fig. 121. Telegraph Relay

very light in order to respond to the action of the current. The clicks of such an armature are not sufficiently loud to be read easily by an operator. Hence, at each station there is introduced a local circuit which contains a local battery, and a second and heavier electromagnet which is called a *sounder*. The electromagnet on

Fig. 122. Standard Morse Local Sounder
Courtesy of Postal Telegraph-Cable Company

the main line is then called the *relay*, see Figs. 121, 122, and 123. The sounder has a very heavy armature, A, Fig. 122, which is so arranged that it clicks both when it is drawn down by its electromagnet against the stop *S* and when it is pushed up again by its spring, on breaking the current, against the stop *t*. The interval which elapses between those two clicks indicates to the operator whether a dot or a dash is sent. The current in the main line simply serves to close and open the circuit in the local battery which operates the sounder, see Fig. 123. The electromagnets of the relay and the sounder differ in that the former consists of many thousand turns of fine wire, usually having a resistance of about 150 ohms, while the latter consists of a few hundred turns of coarse wire having ordinarily a resistance of about 4 ohms.

107. Plan of Telegraph System. The actual arrangement of the various parts of a telegraphic system is shown in Fig. 123. When an operator at Chicago wishes to send a message to New York, he

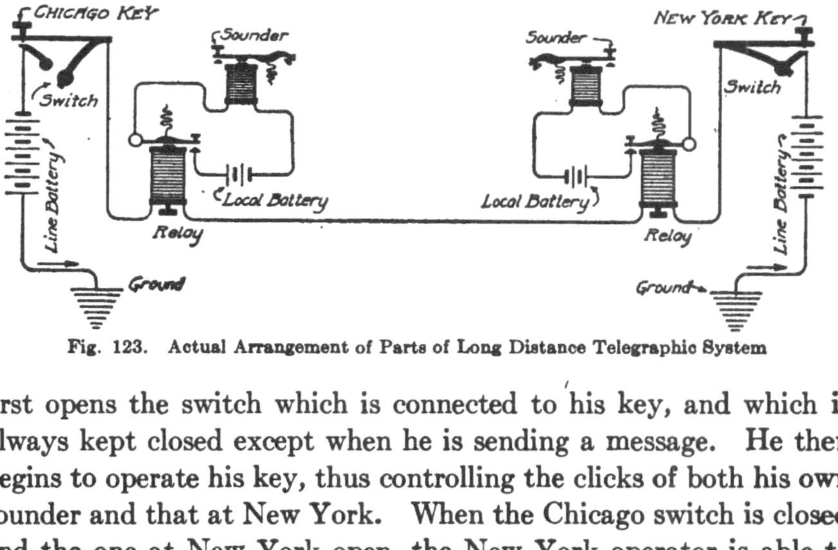

Fig. 123. Actual Arrangement of Parts of Long Distance Telegraphic System

first opens the switch which is connected to his key, and which is always kept closed except when he is sending a message. He then begins to operate his key, thus controlling the clicks of both his own sounder and that at New York. When the Chicago switch is closed and the one at New York open, the New York operator is able to send a message back over the same line. In practice a message is not usually sent as far as from Chicago to New York over a single line, save in the case of transoceanic cables. Instead, it is automatically transferred at, say Cleveland, to a second line which carries it on to Buffalo, where it is again transferred to a third line which carries it on to New York. The transfer is made in precisely the same way as the transfer from the main circuit to the sounder circuit. If, for example, the sounder circuit at Cleveland is lengthened so as to extend to Buffalo, and if the sounder itself is replaced by a relay, called in this case a *repeater*, and the local battery by a main battery, then the sounder circuit has been transformed into a repeater circuit, and all the conditions are met for an automatic transfer of the message at Cleveland to the Cleveland-Buffalo line. There is, of course, no time lost in this automatic transfer.

The necessity of using relays rather than sounders in the line on any long-distance circuit will be clearer after a study of the following examples:

Examples. 1. Ordinary No. 9 telegraph wire has a resistance of 20 ohms to the mile. (a) What current will 100 Daniell cells send through 200 miles of such wire, if there is included at each end of the line, a relay of a resistance of 150 ohms? (Assume for each cell an e.m.f. of 1.05 volts and an internal resistance of 5 ohms.) (b) If the electromagnet in one of the above relays had 10,000 turns of wire on its windings, how many ampere-turns were effective in magnetizing it?

(a) Since the external resistance, 4000 ohms for the line and 300 ohms for the relays, is large in comparison to the internal resistance, the cells should be connected in series. Hence

$$I=\frac{100\times 1.05}{(200\times 20)+(2\times 150)+(100\times 5)}$$

$$=.0219 \qquad \text{Ans. 0.0219 ampere}$$

(b) Ampere-turns = amperes × turns

$$=.0219\times 10{,}000=219$$

$$=219 \text{ ampere-turns}$$

Ans. 219 ampere-turns

2. If sounders having a resistance of 4 ohms each and 600 turns on the electromagnet had been used in place of the relays on the above telegraph line, how many ampere-turns would have been effective in magnetizing their electromagnets?

$$I=\frac{100\times 1.05}{(200\times 20)+(2\times 4)+(100\times 5)}$$

$$=.0233$$

or current = .0233 ampere

ampere-turns = .0233 × 600 = 13.98 Ans. 13.98 ampere-turns

The two examples just given show that the current is nearly as large with relays in the line as with sounders, but with sounders in the line, the ampere-turns are not sufficient to magnetize their electromagnets due to the small number of turns on a sounder, whereas with relays in the line the ampere-turns are sufficient to magnetize the electromagnet of the relay enough to actuate a light armature which operates the local sounder circuit. Hence, we see that on long lines *relays* must be used in the main line rather than *sounders*.

EXAMPLES FOR PRACTICE

1. The plane of a suspended loop of wire is north and south. A current is sent around the loop, passing from north to south on the upper side. When the loop comes to rest, will the face of the loop which was originally toward the east be toward the north?

2. If one looks down on the ends of a U-shaped electromagnet, does the current encircle the two coils in the same direction or in opposite directions? Does it encircle the *N* pole in a clockwise or counter-clockwise direction?

3. Draw a diagram of a short-distance telegraph line where sounders are used without relays.

4. Draw a diagram of a long-distance telegraph line showing how the relays and sounders operate.

5. How would you measure the resistance of the windings on a sounder? On a relay?

6. If a telegraph line between Chicago and Milwaukee (85 miles) is built of No. 9 telegraph wire which has a resistance of 20 ohms per mile, what current will 40 Daniell cells send through the line, if at each end there is connected a relay having a resistance of 140 ohms? (Assume each cell to have an e.m.f. of 1 volt and an internal resistance of 4.5 ohms.) Ans. 0.0185 ampere

7. If in Example 6, sounders of a resistance of 5 ohms each had been used at each end of the line, what current would the 40 Daniell cells have sent through the line? Ans. 0.0212 ampere

8. Find how many ampere-turns were effective in magnetizing the electromagnet of one of the relays in Example 6, if the relay had 10,500 turns of wire on it? Ans. 193+ ampere-turns

9. How many ampere-turns were effective in magnetizing the electromagnet of one of the sounders in Example 7, if the sounder had 600 turns of wire on it? Ans. 12.7 ampere-turns

10. Compare the answers to Examples 8 and 9 and explain why relays, rather than sounders, must be placed "in the line" on long telegraph lines.

108. Other Applications of Electromagnet. The electromagnet is used in so many ways besides those already given that only mention of some of their uses will be made here, the applications being left for the next paper. Telephone instruments, dynamos, motors, induction-coils, automatic operating switches, signalling devices, etc., all utilize the electromagnet in their construction. Strong electromagnets are used extensively in foundries and other places where large masses of iron, rolled plates, steel beams, pig iron, etc., are to be lifted. Fig. 124 shows a large electromagnet employed in the loading of wire scrap.

Fig. 124. Electromagnet Loading Wire Scrap
Courtesy of Link Belt Company, Chicago, Illinois

CHEMICAL EFFECTS OF THE ELECTRIC CURRENT

109. Electroplating. If the tube of Fig. 89 is filled with a solution of copper sulphate ($CuSO_4$), platinum electrodes being used, and the current from one or two dry cells is sent through it, the negative electrode will soon be found to be covered with a bright coat of copper. The reason for this is obvious if one remembers, from the treatment of batteries in Part I, that all metallic ions in solution carry + charges of electricity. In this case, then, the + charged copper ion is repelled by the positive electrode and attracted by the negative electrode where it gives up its charge to the plate and is deposited as metallic copper. Similarly, the − charged SO_4 ion goes to the + plate where it unites with the H_2

of the water forming H_2SO_4, sulphuric acid, and thus frees the oxygen of the water which rises as a gas to the surface. Reversing the current through the U-tube would cause copper to be deposited on the other electrode and remove the copper from the one already plated, since the copper-plated electrode is now the plus electrode and as soon as the SO_4 ions reach it, they unite with the copper and form $CuSO_4$, copper sulphate.

The above experiment illustrates the whole process of electroplating—a process which is used very extensively for obtaining gold- and silver-plated ware; for nickel-plating iron so as to prevent it from rusting; for copper-plating electric-light carbons so as to increase their conductivity, and so on. In commercial work, the positive plate, that is, the plate at which the current enters the bath, is always made from the same metal as that which is to be

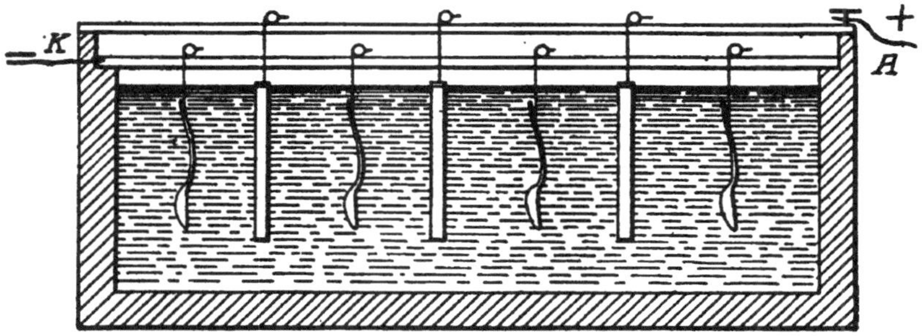

Fig. 125. Section of Silver-Plating Bath

deposited from the solution; for, in this case, the SO_4 or other negative ions dissolve this plate as fast as the metal ions are deposited upon the other. The strength of the solution, therefore, remains unchanged. In effect, the metal is simply taken from one plate and deposited on the other. Fig. 125 represents a silver-plating bath. The bars joined to the anode *A* are of pure silver. The spoons to be plated are connected to the cathode *K*. The solution consists of 500 g. of potassium cyanide and 250 g. of silver cyanide in 10 l. of water.

110. Electrotyping. In the process of electrotyping, the page is first set up in the form of common type. A mold is then taken in wax or gutta-percha. This mold is then coated with powdered graphite to render it a conductor, after which it is ready to be suspended as the cathode in a copper-plating bath, the anode being a plate of pure copper and the liquid a solution of copper sulphate. When a sheet of copper as thick as a visiting card has been deposited on the mold, the latter is removed and the wax replaced by a type-metal backing, to give rigidity to the copper films. From such a plate, as many as a hundred thousand impressions may be made. Practically all books which run through large editions are printed from such electrotypes. The printed matter which you are now reading was printed from such electrotypes, and yet the printing is so distinct that the average reader would naturally think it had been printed from the actual type used in setting up the original form for the printed page.

111. Refining of Metals. If the solution consists of pure copper sulphate, it is not necessary that the anode be of chemically pure copper in order to obtain a pure copper deposit on the cathode. *Electrolytic* copper, which is the purest copper on the market, is obtained as follows: The unrefined copper is used as an anode. As it is eaten up, the impurities contained in it fall as a residue to the bottom of the tank and pure copper is deposited on the cathode by the current. This method is also extensively used in the refining of metals other than copper.

112. Chemical Method of Measuring Current. An ampere will deposit in an hour 1.181 g. of copper, 1.203 g. of zinc, etc. This fact is made use of in calibrating fine ammeters, since it is possible to compute with great accuracy the strength of a current which has deposited a given weight of metal in a known length of time. This method of calibrating an ammeter has been treated in Section 82.

113. Storage Batteries. A simple storage cell can be made with two six-inch by eight-inch lead plates, screwed to a half-inch strip of some insulating material, as in Fig. 126, and immersed in a solution consisting of one part of sulphuric acid to ten parts of water. Let a current from two bichromate cells *C* be sent through this arrangement, an ammeter *A* or any low-resistance galvanometer being inserted in the circuit. As the current flows, hydrogen bubbles will be seen to rise from the cathode (the plate at which the current leaves the solution), while the positive plate, or anode, will begin to turn dark brown. At the same time the reading of the ammeter will be found to decrease rapidly. The brown coating is a compound of lead and oxygen called lead peroxide, PbO_2, which is formed by the action upon the plate of the oxygen which is liberated precisely as in the experiment on the electrolysis of water, Section 81. Now let the batteries be removed from the circuit by opening the key K_1, and let an electric bell B be inserted in their place by closing the key K_2. The bell will ring and the ammeter A will indicate a current flowing in a direction opposite to that of the original current. This current will rapidly decrease as the energy which was stored in the cell by the original current is expended in ringing the bell.

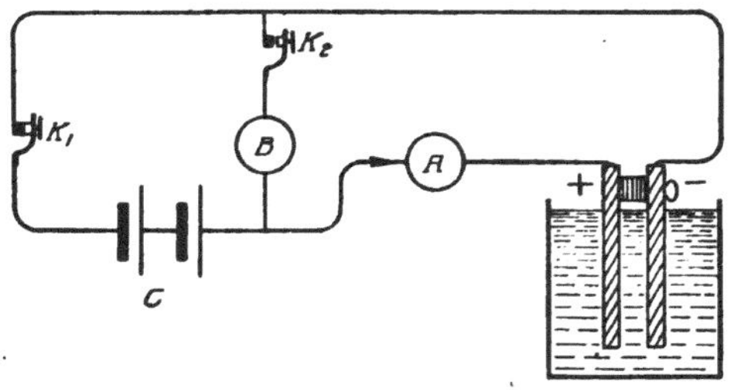

Fig. 126. Circuit Showing Simple Method of Making Storage Battery

This experiment illustrates the principle of the *storage battery.* Properly speaking, there has been no storage of *electricity,* but only a storage of *chemical energy.*

Two similar lead plates have been changed by the action of the current into two dissimilar plates, one of lead, and one of lead peroxide. In other words, an ordinary galvanic cell has been formed, for any two dissimilar metals in an electrolyte constitute a primary galvanic cell. In this case the lead peroxide plate corresponds to the copper of an ordinary cell, and the lead plate to the zinc. This cell tends to create a current opposite in direction to that of the charging current; i.e., its e.m.f. pushes back against the e.m.f. of the charging cells. It was for this reason that the ammeter reading fell. When the charging current is removed, this *primary* galvanic cell will furnish a current until the thin coating of peroxide is used up. The only important difference between a commercial storage cell, Fig. 127, and the one which we have here used is that the former is provided in the making with a much thicker coat of the "active material", lead peroxide, on the positive

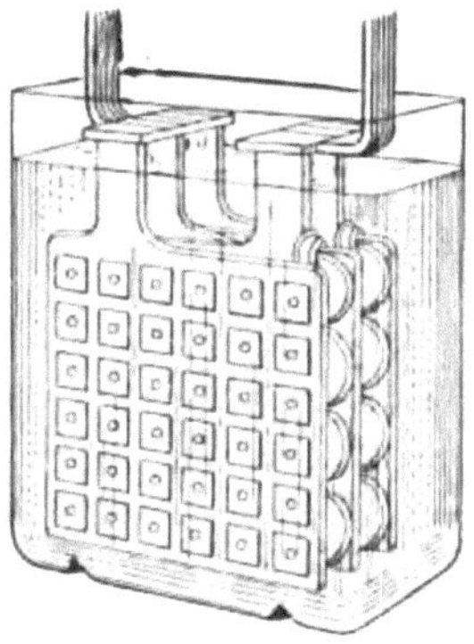

Fig. 127. Typical Storage Battery Plates

plate and a porous spongy lead on the negative, than can be formed by a single charging such as we used. This material is pressed into interstices in the plates, as shown in Fig. 127. The e.m.f. of the storage cell is about 2 volts. Since the plates are always very close together and may be given any desired size, the internal resistance is usually small, so that the currents furnished may be very large. They are sometimes as high as several thousand amperes when many files of cells are connected to form a battery.

The usual efficiency of the storage cell is about 75 per cent, i.e., but three-fourths as much electrical energy can be obtained from it as is put into it.

114. Uses of Storage Battery. The storage battery is probably most often thought of in connection with its use in electric automobiles for power and lighting, and its use in gasoline-driven cars for lighting, ignition, and self-starting devices. Its use, however, in electric power plants is fully as important. In some power plants, the generators are not large enough to supply the highest demand for current during the day, at which time the storage batteries are used to furnish part of the current, the storage batteries being charged by the generator during that portion of the day when the load is light. In some of the small power plants in small cities, the current required for use in the daytime is so small that the generator is not run at all, the power being supplied, during this time of light load, by the storage batteries.

115. Edison Storage Battery. The new Edison storage battery has about twice the output for the same weight as the lead storage battery and is therefore eminently suited for traction purposes. The positive plate is made of nickel oxide, the negative is pure iron in a steel frame, and the electrolyte is a solution of caustic potash. It has an e.m.f. of about 1.2 volts, and an internal resistance somewhat higher than that of the lead cell and is consequently not quite as efficient. Whether or not it will supplant the lead storage battery is difficult to predict.

HEATING EFFECTS OF ELECTRIC CURRENT

116. Heat Developed in Wire by Electric Current. Let the terminals of two or three dry cells in series be touched to a piece of No. 40 iron or German-silver wire and the length of wire between these terminals shortened to one-fourth inch or less. The wire will be heated to incandescence and probably melted.

The experiment shows that just as in the charging of a storage battery, the energy of the electric current was transformed into the energy of chemical separation, so here in the passage of the current through the wire, the energy of the electric current is transformed into heat energy.

117. Energy Relations of Electric Current. If the energy expended on a water turbine were measured, it would be found to equal the quantity of water passing through the turbine, multiplied by the difference in level through which the water falls. In just the same way, it is found that when a current of electricity passes through a conductor, the energy expended is equal to the quantity of electricity passing, multiplied by the difference in potential between the ends of the conductor. If the quantity of electricity is expressed in coulombs and the p.d. in volts, the energy is given in joules, and we have

$$\text{volts} \times \text{coulombs} = \text{joules}$$

Since the number of coulombs is equal to the number of amperes of current multiplied by the number of seconds

$$\text{volts} \times \text{amperes} \times \text{seconds} = \text{joules}$$

But a watt is defined as a joule per second, see Section 94. Hence, the energy expended per second by the current, that is, the *power* of the current, is given by the formula

$$\text{volts} \times \text{amperes} = \text{watts}$$

118. Calories of Heat Developed in a Wire. The electrical energy may be expended in a variety of ways when a current flows between points of given p.d. For example, it may be spent in producing chemical separation, as in the charging of a storage cell; it may be spent in doing mechanical work, as is the case when the current flows through an electric motor; or it may be spent wholly in heating the wire, as was the case in the experiment of Section 116. It will always be expended in the latter way when no chemical or mechanical changes are produced by it. The number of calories of heat produced per second in the wire of the last experiment is found, then, by multiplying the number of joules expended by the current per second, by the heat equivalent of the joule in calories, that is, .24 calorie, since 1 calorie is 4.2 joules. Therefore when all of the electrical energy of a current is transformed into heat energy, we have

$$\text{calories per second} = \text{volts} \times \text{amperes} \times .24$$

The total number of calories H developed in t seconds will be given by

$$H = \text{p.d.} \times C \times t \times .24$$

Thus a current of 10 amperes flowing in a wire whose terminals are at a potential difference of 12 volts will develop in 5 minutes $10 \times 12 \times 300 \times .24 = 8640$ calories.

Since by Ohm's law p.d. $= C \times R$, we have, by substituting CR for p.d. in the above formula

$$H = C^2 R \times t \times .24$$

or, *the heat generated in a conductor is proportional to the time, to the resistance, and to the square of the current.* This is known as Joule's law, from the name of the man who first announced it as the result of his experimental researches.

119. Incandescent Lamps. The ordinary incandescent lamp consists of a carbon filament heated to incandescence by an electric current, Fig. 128. Since the carbon would burn instantly in air, the filament is placed in a highly exhausted glass bulb. Even then it disintegrates slowly. The normal life of a 16-candle-power lamp filament is from 1,000 to 2,000 working hours. The filament is made by carbonizing a special form of cotton thread. The ends of the carbonized thread are attached to platinum wires which are sealed into the glass walls of the bulb, and which make contact, one with the base of the socket and the other with its rim, these being electrodes through which the current enters and leaves the lamp.

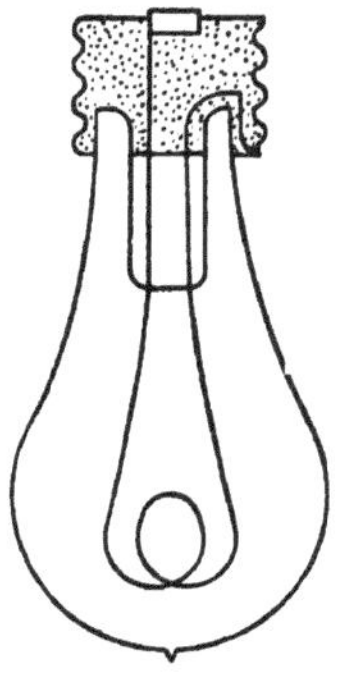

Fig. 128. Simple Carbon Filament Incandescent Lamp

The ordinary 16-candle-power lamp is most commonly run on a circuit which maintains a potential difference of either 110 or 220 volts between the terminals of the lamp. In the former case, the lamp carries about .5 ampere of current, and in the latter case, about .25 ampere. It will be seen from these figures that the rate of consumption of energy is about 3.4 watts per candle-power.

A customer usually pays for his light by the "kilowatt hour", a kilowatt hour being the energy furnished in one hour by a current whose rate of expenditure of energy is one kilowatt. Thus the rate at which energy is consumed by a 16-candle-power lamp is $110 \times .5 = 55$ watts, or .055 kilowatt. One such lamp running for ten hours would therefore consume 550 watt hours of energy or .55 kilowatt hour.

At the present time tungsten filaments are being used very largely for incandescent lamps. They are nearly three times as efficient as the carbon lamps, the "Mazda" form taking but 1.25 watts per candle. This is because they can be operated at much higher temperatures than carbons can. They are, however, somewhat more expensive.

120. Arc Light. When two carbon rods are placed end to end in the circuit of a powerful electric generator, the carbon about the point of contact is heated red hot. If, then, the ends of the carbon rods are separated one-fourth inch or so, the current will still continue to flow, for a conducting layer of incandescent vapor called an *electric arc* is produced between the poles. The appearance of the arc is shown in Fig. 129. At the + pole a hollow, or crater, is formed in the carbon, while the − carbon becomes cone shaped, as in the figure. The carbons are consumed at the rate of about an inch an hour, the + carbon wasting away about twice as fast as the − one. The light comes chiefly from the + crater, where the temperature is about 3,800° C., the highest attainable by man. All known substances are volatilized in the electric arc.

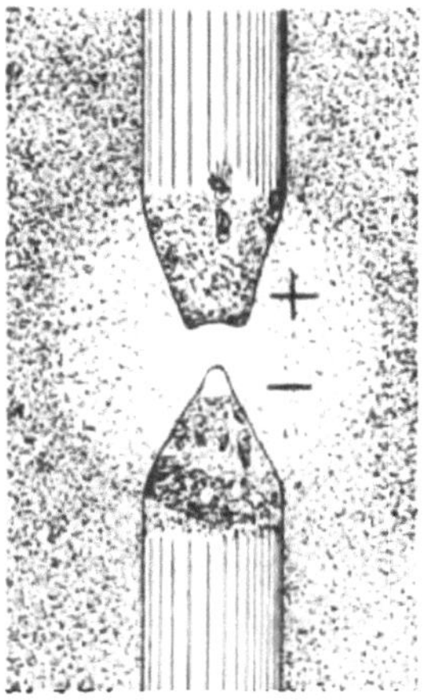

Fig. 129. Arc Formation in Arc Lamp

The ordinary arc requires a current of 10 amperes and a p.d. between its terminals of about 50 volts. Such a lamp produces about 500* candle-power, and therefore consumes energy at the

*This is the so-called "mean spherical" candle-power. The candle-power in the direction of maximum illumination is from 1,000 to 1,200.

rate of about 1 watt per candle-power. This makes an arc light about 3.5 times as efficient as an incandescent light. The recently invented *flaming arc,* produced between carbons which have a composite core consisting of carbon, lime, magnesia, silica, or other light-giving minerals, sometimes reaches an efficiency as high as .27 watt per candle-power.

121. Arc-Light Automatic Feed. Since the two carbons of the arc gradually waste away, they would soon become so far separated that the arc could not longer be maintained were it not for an automatic feeding device which keeps the distance between the carbon tips very nearly constant. Fig. 130 shows the essential features of one form of this device. When no current is flowing through the lamp, gravity holds the carbon tips at *e* together; but as soon as the current is thrown on, it energizes the low-resistance electromagnet *M*, which is in series with the carbons. This draws down the iron plunger *c*, which acts upon the lever *L*, and "strikes the arc" at *e*. But the introduction of the resistance of the arc into the circuit *sABMt* raises the p.d. between *s* and *t* and thus causes an appreciable current to flow through the high-resistance magnet *N*, which is shunted across this circuit. This tends to *raise* the plunger *c* and thus to shorten the arc. There is thus one particular length of arc for which equilibrium exists between the effects of the series magnet *M* and the shunt magnet *N*. This length the lamp automatically maintains. The magnet *O* is the so-called "cut-out" inserted so that, if the lamp gets out of order and the arc burns out, a current at once flows through *N* and *O* of a strength sufficient to close the contact points at *r* and thus permit the main current to flow on to the next lamp over the path *PrRQ*.

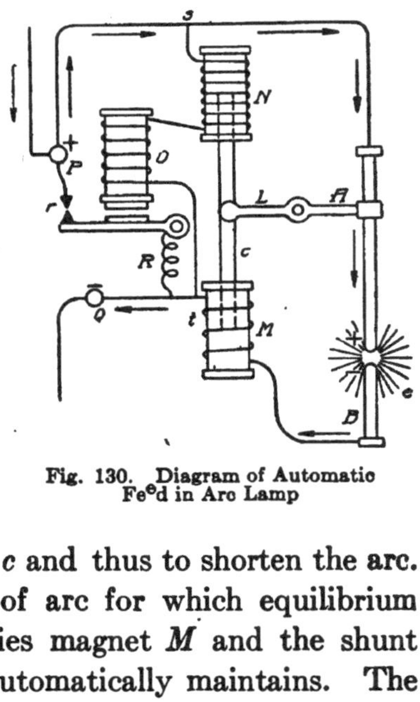

Fig. 130. Diagram of Automatic Feed in Arc Lamp

122. Cooper-Hewitt Mercury Lamp. The Cooper-Hewitt mercury lamp, Fig. 131, is the most efficient of all electric lights, unless it be the flaming arc. It differs from the arc lamp in that the incandescent body is a long column of mercury vapor instead of an incandescent solid. The lamp consists of an exhausted tube three or four feet long, the positive electrode at the top being a plate of iron, and the negative electrode at the bottom a small quantity of mercury. Under a sufficient difference of potential between these terminals, a long mercury-vapor arc is formed which stretches from terminal to terminal in the tube. This arc emits a very brilliant light, but it is almost entirely wanting in red rays. The efficiency of the lamp is very high, since it requires but .3 watt per candle-power. It is rapidly finding important commercial uses, especially in photography. The chief objection to it arises from the fact that, on account of the absence of red rays, the light gives objects an unnatural color.

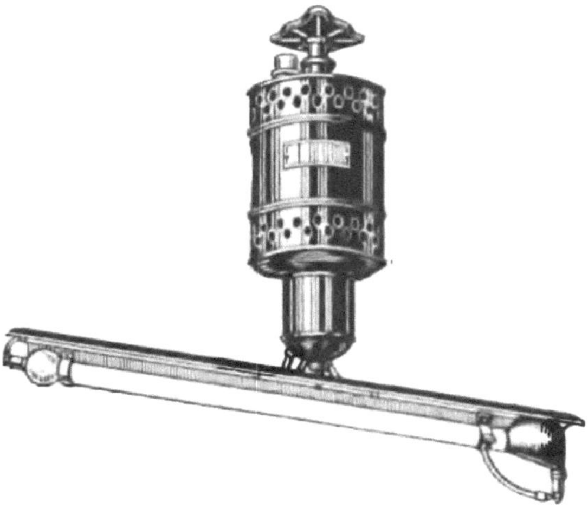

Fig. 131. Cooper-Hewitt Mercury Lamp

123. Household and Artisans' Devices. The application of electric currents to heating in recent years has taken a very prominent place in a multitude of devices used in the trades and in the home. Our highest temperatures are obtained in the electric arc and the electric furnace. Electric disk stoves, teakettles, egg-cookers, toasters, chafing-dishes, percolators, grills, water-heaters, immersion coils, flatirons, curling-irons, mangles, foot-warmers, radiators, shaving-mugs, milk-warmers, and sterilizers are some of the household conveniences. Electric soldering-pots, soldering-irons, glue-cookers, gluepots for cabinet-makers, melting-pots for lead alloys, instantaneous heaters for soda fountains, mangles, fluting-irons, automobile-tire vulcanizers, and electric welders are some of the commercial applications of heating by the electric current.

The designer and the producer of these useful appliances are interested in increasing the percentage of the energy of the electric current which is transformed into heat available for the intended purpose. The ratio of this available heat energy, or output, to the electric energy, or input, for any of these devices is its efficiency.

The consumer or user of these appliances is interested, not only in their efficiency and cost of operation, but also in their convenience.

124. Efficiencies of Household Water-Heaters. The most commonly used electric devices for heating small quantities of water in the home are the disk stove, teakettle, and immersion heater, similar to those shown in Figs. 132, 133, and 134.

Fig. 132. Simple Electric Stove

To find the efficiency of the heater use about 500 grams of water which is 12° or 15° C. below room temperature; connect the heater, as shown in Fig. 135, in series with an ammeter in the line and a voltmeter across the line. Then turn on the current and stir the water continually, observing the exact time when the water attains a temperature which is 10° C. below room temperature, and again when the water attains

Fig. 133. Electric Teapot

Fig. 134. Electric Immersion Heater

a temperature which is 10° C. above room temperature. The useful heat obtained, the output, during the observed time, is then expressed in calories by multiplying the weight of the water in grams by its rise in temperature in degrees centigrade. The corresponding electrical energy used, the input, also expressed in calories, is given by any one of the following equations:

$$H = .24C^2 \times R \times t$$
$$H = .24C \times E \times t$$
$$H = .24 \times \text{watts} \times \text{seconds}$$

The efficiency, as in all machines, is then given by the equation

$$\text{Efficiency} = \frac{\text{output}}{\text{input}}$$

which in this case becomes

$$\text{Efficiency} = \frac{\text{grams of water} \times \text{degrees centigrade rise in temperature}}{.24 \times \text{watts} \times \text{seconds}}$$

Example. A certain electric teakettle carrying 4.2 amperes on a 110-volt circuit, which will heat 500 grams of water from 14° C. to 34° C. in 2 minutes has an efficiency of

$$\frac{500 \times 20}{.24 \times 4.2 \times 110 \times 120} = .75+, \text{ or } 75+ \text{ per cent}$$

Ans. .75+, or 75+ per cent

In exactly the same way the efficiency of heating water in an ordinary teakettle with the disk stove of Fig. 132 may be found.

Fig. 135. Typical Circuit for Investigating Efficiency of Electric Heating Apparatus

In this case, however, the heating unit is not a part of the teakettle so that a much larger portion of the heat is lost and consequently its efficiency is much lower, being about 40 per cent. If the immersion heater of Fig. 134 is used, the heating unit is then completely surrounded by water so that only a very small portion of the heat is lost, and its efficiency is therefore very high, being about 95 per cent when used with an ordinary dish to contain the water.

125. Protection of Circuits against Overheating. Fuses and Circuit Breakers. If an electric-lighting circuit were to become short-circuited accidentally, for example, by the wearing off of the insulation on the drop-cord leading to an electric lamp, the current flowing would be so large that either the wiring in the house, or the drop-cord, would be melted at some point and the heat would in many cases be sufficient to set fire to the building. To protect circuits against accidents of this sort, fuses are introduced into the circuit. The fuse is made of a strip or a wire of alloy of such a current-carrying capacity that it will melt from the heat produced by a current which is much in excess of that marked on the fuse by the maker. Thus the circuit is automatically broken in the "fuse box" and no harm can result to other parts of the circuit. Fig. 136 shows the simple wire fuse, Fig. 137 the "cartridge fuse", which is protected by an asbestos covering, and Fig. 138 the "plug fuse", which is placed inside a hollow porcelain container and is then screwed into a socket similar to a lamp socket.

Fig. 136. Simple Wire Fuse

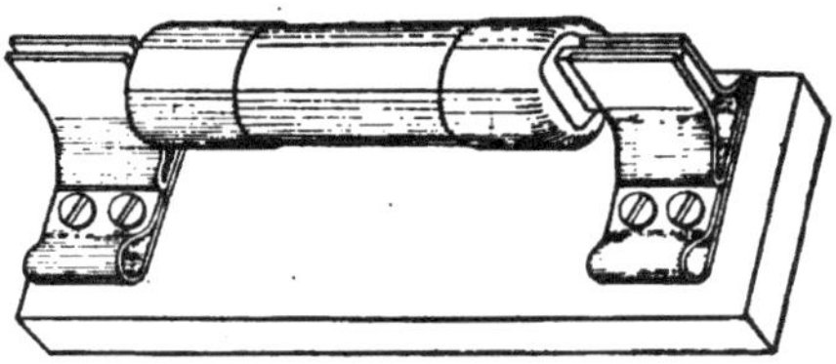

Fig. 137. Cartridge Fuse

Where large currents are being used in power stations, and for running street cars, etc., protection against excessive currents is secured by means of the circuit breaker, which is nothing but an electromagnet about which the current passes and which will be strong enough to open a switch, thus automatically breaking the circuit when the current rises to a certain maximum load. A circuit breaker on a street car will prevent a motorman from sending enough current through the motor, when starting, to heat the windings of the armature and field coils, up to a temperature which would burn the insulation of the wires and completely ruin the motor.

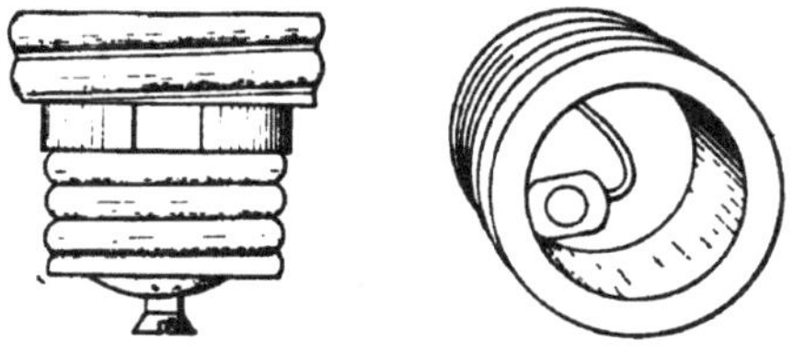

Fig. 138. Standard Plug Fuse

EXAMPLES FOR PRACTICE

1. An electric foot-warmer carries 1.2 amperes on a 110-volt circuit. What does it cost per hour to run it at 8 cents per kilowatt hour for the electricity? Ans. 1.06 cents

2. An electric flatiron carries 4.4 amperes on a 110-volt circuit. What does it cost per hour to run it at 8 cents per kilowatt hour for the electricity? Ans. 3.87 cents

3. How many calories of heat are developed per hour by the flatiron in Example 2? Ans. 418,176 calories

4. How long will it take an electric teakettle which has an efficiency of 70 per cent and which carries a current of 4 amperes on a 110-volt circuit, to heat 2 liters (2.11 quarts or 2,000 grams) from 10° C. to 80° C.? Ans. 31 min. 34 sec.

5. How long would it take the heater of Example 4 to heat 500 grams of water (about 1 pint) through the same range of temperature? Ans. 7 min. 54 sec.

6. Why is electricity not more generally used for water?

ELEMENTS OF ELECTRICITY AND MAGNETISM

PART III

INDUCED CURRENTS AND ELECTRIC POWER

126. Faraday's Discovery. The production of currents of electricity by electromagnetic induction was first accomplished in 1831 by Michael Faraday (1791–1867). At the age of thirteen, Faraday was apprenticed to a London bookbinder with whom he worked nine years; in 1813, he obtained a position in Sir Humphry Davy's laboratory at the Royal Institution, of which he became director in 1825; in 1831, he discovered electromagnetic induction, the climax to a series of experiments upon which he had been working for six years in an attempt to find the laws of the actions of magnets on currents; later, he made the first dynamo; in 1833, he announced the laws of electrolysis, now known as Faraday's laws; in his honor, the practical unit of electricity, the *farad*, has been named. The great scientist, Tyndall, pronounced Faraday's discovery of electromagnetic induction to be the greatest experimental result ever obtained. Its discovery made possible the modern age of electricity, for without the dynamo we should have to depend upon batteries for all of our electric currents, the cost of which, when produced by primary cells, would be prohibitive on account of the high cost of zinc. The cost of operating storage batteries for electric vehicles would also prohibit their use if charged by primary batteries, but since these are charged by induced currents produced by the dynamo, the cost is enormously reduced. Without dynamo electric machinery, in fact, we would not be enjoying our modern systems of electric lighting, our shops equipped with motor-driven machinery, our motor vehicles, or the thousand and one electrical household conveniences or artisan's appliances which we have today.

127. Induction of Currents by Magnets. The principle of electromagnetic induction may be simply shown by a home-made apparatus, Fig. 139. Wind 400 or 500 turns of No. 22 copper wire into a coil *C* about two and one-half inches in diameter. Connect this coil with a lecture-table galvanometer, Fig. 150, or to a simple detector, Fig. 139. The coil of the detector may be made with about 100 turns of the same wire, and should be placed with its plane in the magnetic meridian; its magnetic needle may be made of any magnetized piece of steel suspended by one strand of silk-twist thread, or it may be an ordinary compass. Thrust the coil *C* suddenly over the *N* pole of a strong horseshoe magnet. The deflection of the needle of the detector will indicate a momentary current flowing through the coil of the detector. If the coil *C* is held stationary over the magnet, the needle will be found to come to rest in its natural position. If the coil *C* be removed suddenly from the

Fig. 139. Illustrating Principle of Electromagnetic Induction

pole, the needle will move in a direction opposite to that of its first deflection, showing that a reverse current is now being generated in the coil.

We learn, therefore, *that a current of electricity may be induced in a conductor by causing the latter to move through a magnetic field,* while a magnet has no such influence upon a conductor which is at rest with respect to the field.

128. Direction of Induced Current. *Lenz's Law.* In order to find the direction of the currents induced in the coil *C*, we may apply a very small p.d. from a galvanic cell to the terminals of the coil of the detector, and note the direction in which the needle moves when the current encircles the detector coil in the given direction. (This could be determined easily without a voltaic cell by applying either Ampère's right-hand rule or the helix rule to the current flowing in the detector coil.) This will at once show in what direction the current was flowing in the coil *C* when it was being thrust over the *N* pole. By a simple application to *C* of the right-hand helix rule, Section 101, we can then tell which was the *N* and which the *S* face of the coil when the induced current was flowing through it. In this way it will be found that if the coil was being moved past the *N* pole of the magnet, the current induced in it was in such a direction as to make the lower face of the coil an *N* pole during the downward motion and an *S* pole during the upward motion. In the first case, the repulsion of the *N* pole of the magnet and the *N* pole of the coil *opposed* the motion of the coil while it was moving from *a* to *b;* and the attraction of the *N* pole of the magnet and the *S* pole of the coil *opposed* the motion while it was moving from *b* to *c*. In the second case, when the coil *C* is removed from the magnet, the repulsion of the two *N* poles *opposed* the upward motion from *c* to *b*, and the attraction between the *N* pole of the magnet and the *S* pole of the coil *opposed* the upward motion from *b* to *a*. *In every case, therefore, the motion is made against an opposing force.*

From these and similar experiments, we arrive at the following law: *Whenever a current is induced by the relative motion of a magnetic field and a conductor, the direction of the induced current is always such as to set up a magnetic field which opposes the motion.* This law is known as Lenz's law. It might have been predicted at once from the principle of the conservation of energy; for this principle tells us that since an electric current possesses energy, such a current can appear only through the expenditure of mechanical work or of some other form of energy. Thus if the induced currents in the coil *C* had been in such a direction as to set up a magnetic field which aided the motion instead of opposing it, the coil *C* would move without external force being applied to it and we would have energy in the form of electric current produced without any expenditure of energy, or in other words, we would have perpetual motion. Thus again a law of Physics points to the fallacy of the perpetual-motion idea.

129. Conditions Necessary for an Induced E.M.F. If the coil *C* be held in the position shown in Fig. 140, and then be moved back and forth *parallel* to the magnetic field, that is, parallel to the line *NS*, no current will be induced. If a strong electromagnet is available, this experiment is more instructive if performed, not with a coil, as in Fig. 140, but with a straight rod, Fig. 141, to the ends of which are attached wires leading to a galvanometer. Whenever the rod moves *parallel* to the lines of magnetic force, there will be no induced e.m.f. and therefore no deflection of the galvanometer, but whenever it moves across the lines, the galvanometer needle will at once move.

By experiments of this sort, it is found that an e.m.f. is induced in a coil only *when the motion takes place in such a way as to change the total number of magnetic lines of force which are inclosed by the coil.* Or, to state this rule in more general form, *an e.m.f. is induced in any element of a conductor when, and only when, that element is moving in such a way as to cut magnetic lines of force.*

It will be noticed that the first statement of the rule is included in the second, for whenever the number of lines of force which pass

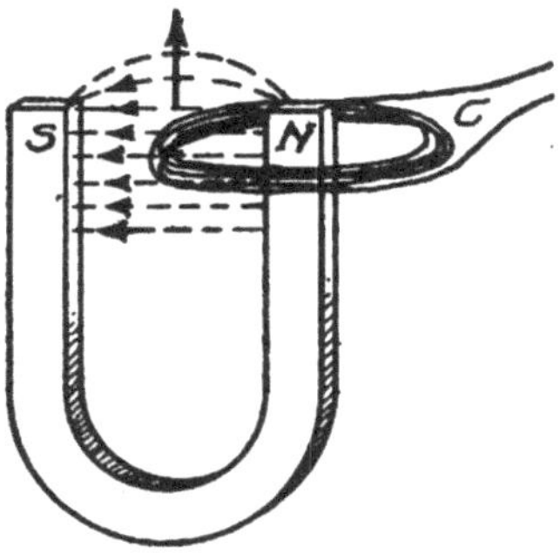

Fig. 140. Conductor Cutting Lines of Force to Induce an E.M F.

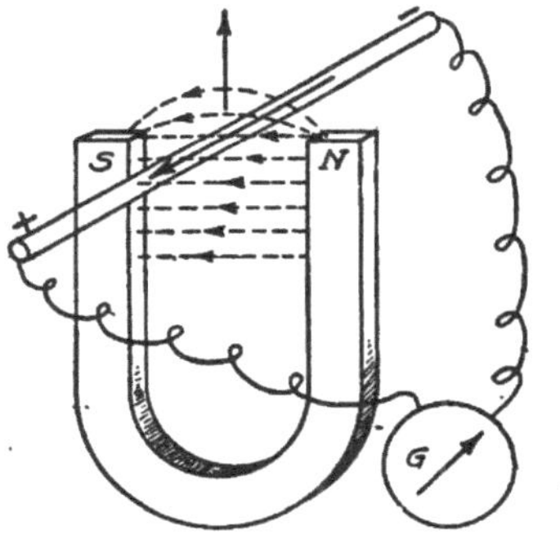

Fig. 141 E M F Induced by Straight Conductor Cutting Magnetic Lines

through a coil changes, some lines of force must cut across the coil from the inside to the outside, or vice versa.

In stating these rules, we have used the expression *induced e.m.f.* instead of *induced current* for the reason that whether or not a continuous current flows in a conductor in which an e.m.f. (i.e., a pressure tending to produce a current) exists, depends simply on whether or not the conductor is a portion of a closed electrical circuit. In our experiment, the portion of the wire in which the e.m.f. was being generated by its passage across the lines of force running from N to S was a part of such a closed circuit, and hence a current resulted. If we had moved a straight conductor like that shown in Fig. 142, the e.m.f. would have been induced precisely as before; but since the circuit would then have been open, the only effect of this e.m.f. would have been to establish a p.d. between the ends of the wire, i.e., to cause a positive charge to appear at one of its ends and a negative charge at the other, in precisely the same way that the e.m.f. of a battery causes positive and negative charges to appear on the terminals of the battery when it is on open circuit.

130. Strength of the Induced E.M.F. The strength of an induced e.m.f. is found to depend simply upon *the number of lines of force cut per second by the conductor*, or, in the case of a coil, upon the *rate of change* in the number of lines of force which pass through the coil. The strength of the current which flows is then given by Ohm's law, i.e., it is equal to the induced e.m.f. divided by the resistance of the circuit. The number of lines of force which the conductor cuts per second may always be determined if we know the velocity of the conductor and the strength of the magnetic field through which it moves.*

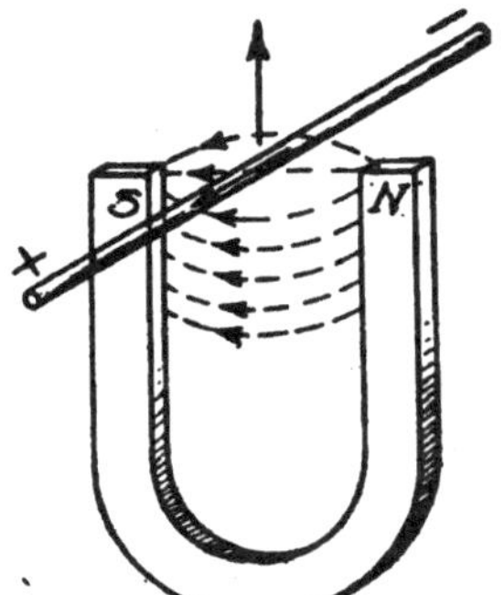

Fig 142. Direction of Induced Currents for Given Direction of Motion

In a conductor which is cutting lines at the rate of 100,000,000 lines per second, there is an induced electromotive force of exactly one volt.

131. Dynamo, or "Right-Hand", Rule. Since Fig. 139 shows that reversing the direction in which a conductor is cutting lines of force also reverses the direction of the induced electromotive force, it is clear that a fixed relation must exist between these two directions and the direction of the magnetic lines. What this relation is may be obtained easily from Lenz's law. When the conductor was moving upward, Fig. 141, the current flowed in such a direction as to oppose the motion, that is, so as to make the lower face of the coil an *S* pole. This means that in the portion of the conductor between *N* and *S* where the e.m.f. was being generated,

*A magnetic pole of unit strength is by definition a pole, which, when placed at a distance of one centimeter from an exactly similar pole, repels it with a force of one dyne (about one thousandth of a gram, or $\frac{1}{28000}$ of an ounce). A magnetic field of unit strength is, by definition, a field in which a unit pole is acted upon by a force of one dyne. Hence, if a unit pole is found in a given field to be acted upon by a force of one thousand dynes, we say that the field strength is one thousand units. Now, it is customary to represent a magnetic field by drawing as many lines per square centimeter taken at right angles to the direction of the field as the field has units of strength. Thus, a field of unit strength is said to contain one line per square centimeter, a field of the strength of a thousand units, a thousand lines per square centimeter, etc. The magnetic fields used in powerful dynamos will have sometimes as high as 20,000 lines per square centimeter.

its direction was from back to front, that is, toward the reader, as the arrow indicates, Fig. 141. We therefore set up the following rule, which is found to apply in every case:

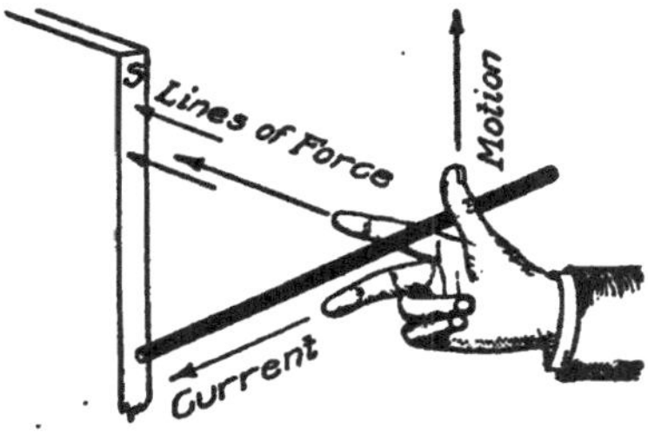

Fig. 143. Illustrating the Dynamo, or Right-Hand, Rule

If the forefinger of the **right hand** *points in the direction of the magnetic lines, Fig. 143, and the thumb in the direction in which the conductor is cutting these lines, then the middle finger, held at right angles to both thumb and forefinger, will point in the direction of the induced current.*

This rule is known as the *dynamo rule.*

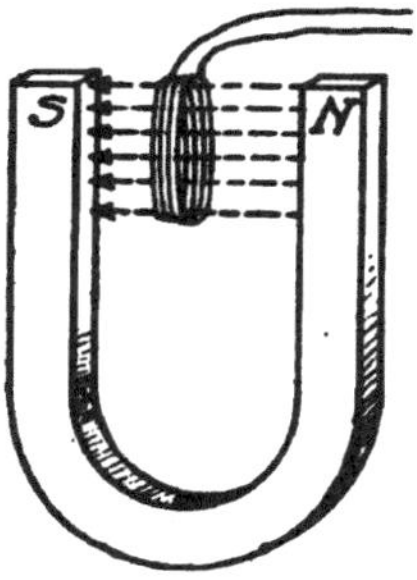

Fig. 144. Position of Coil for Weakest Induced Current

132. Principle of the Dynamo. A dynamo is essentially nothing but a coil of wire rotating continuously between the poles of a magnet. Thus, suppose that starting with the coil in the position shown in Fig. 144, it be rotated through 180 degrees from left to right as one looks down upon it. During the first half of the revolution, the wires on the right side of the loop are cutting the lines of force while moving toward the reader, while the lines on the left side are cutting the same lines while moving away from the reader. Hence, by applying the dynamo rule, we find that a current is being generated which flows down on the right side of the coil and up on the left side. It will be seen that both currents flow around the coil in the same direction. The induced current is strongest when the coil is in the position shown in Fig. 145, because there the lines of force are being cut most rapidly. Just as the coil is being moved into or out of the position shown in Fig. 144, it is moving *parallel* to the lines of force and hence no current is induced, since no lines of force are being cut. As the coil is now moved through the last 180 degrees of a complete revolution, both sides are cutting the same lines of force as before, but they are cutting them while moving in a direction

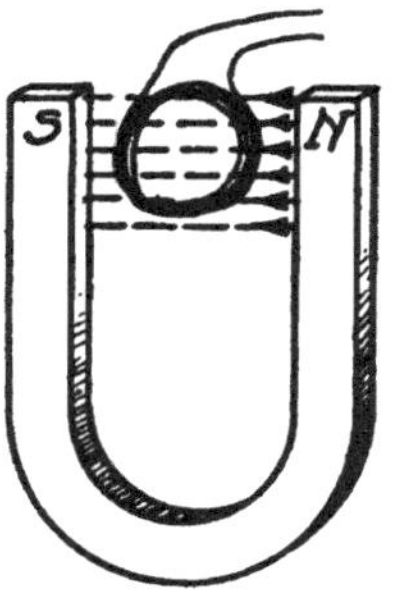

Fig. 145. Position of Coil for Strongest Induced Current

opposite to that in which they were first moving, hence the current generated during this last half is opposite in direction to that of the first half. If the coil is continuously rotated in the field, therefore, an alternating current is set up in it, which reverses direction every time the coil passes through the position shown in Fig. 144. This is the essential principle of the alternating-current dynamo. The direct-current dynamo differs from the alternating-current dynamo, only in that a so-called *commutator* is used for the purpose of changing the direction of the current in the external circuit, every time the coil passes through the position shown in Fig. 144, so that the current always flows in the same direction through this external portion of the circuit in spite of the fact that in the rotating coil it changes direction every half-revolution.

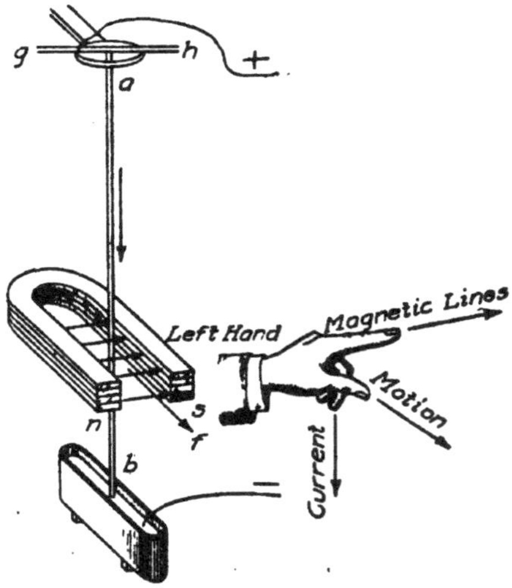

Fig 146. Illustrating Principle of Electric Motor

133. Principle of the Electric Motor. If a vertical wire *ab* is made to pass between the poles of a magnet in the manner shown in Fig. 146, and the current from an outside source—for example, a Leclanché cell—sent through it from *a* to *b*, the wire *ab* will be found to move through the mercury, into which its lower end dips, in the direction indicated by the arrow *f*, namely, at right angles to the direction of the lines of magnetic force. If the direction of the current in *ab* is reversed, the direction of the motion of the wire will be found to be reversed also. This experiment shows that *a wire carrying a current in a magnetic field tends to move in a direction at right angles both to the direction of the field and to the direction of the current.* The experiment illustrates the essential principle of the electric motor.

Motor, or "Left-Hand", Rule. The relation between the direction of the magnetic lines, the direction of the current, and the direction of the force, is often remembered by means of the *motor rule.*

It differs from the dynamo rule, only in that it is applied to the fingers of the *left hand* instead of to those of the right.

Let the forefinger of the **left hand** *point in the direction of the magnetic lines of force and the middle finger in the direction of the current sent through the wire; the thumb will then point in the direction of the mechanical force acting to move the wire.* (See Fig. 146.)

In practice, the motor does not differ in construction at all from the dynamo. Thus, if a current is sent into the right side of the coil, Fig. 145, and out of the left side, the wires on the left side of the coil will be seen, by an application of the motor rule, to be urged toward the reader, while the wires on the right side are urged away from the reader; hence the coil begins to rotate. After it has rotated through the position shown in Fig. 144, if the direction in which the current flows through it were not changed, it would be urged to rotate back to the position of Fig. 144; but in the actual motor, at the instant at which the coil passes through the position shown in Fig. 144, the commutator reverses the direction of the current as it enters the coil; hence, the coil is always impelled to rotate in the same direction.

134. Faraday's Conception of the Motor and the Dynamo Principles. Faraday thought of the lines of force as acting like stretched rubber bands, a conception which sometimes enables one to picture to himself more clearly, for example, just why a motor runs.

If the field of a magnet alone is represented by Fig. 147 and that due to the current alone by Fig. 148, then the resultant field when the current-bearing wire is placed between the poles of the magnet is that shown in Fig. 149, for the strength of the field above the wire is now the sum of the two separate fields, while the strength below it is their difference. If, now, we think of the lines of force as acting like stretched rubber bands, as did Faraday, then we say that the wire in Fig. 149 would be pushed down. In the case of the dynamo principle, when a wire is moved through a magnetic field, the current induced in it must be in such a direction as to *oppose* the motion; therefore, *the induced current will be in such a direction as to increase the number of lines on the side toward which it is moving.*

EXAMPLES FOR PRACTICE

1. Under what conditions may an electric current be produced by a magnet?

2. State Lenz's law, and show by the principle of the conservation of energy that this law must be true.

3. A current is flowing up in a vertical wire. In what direction will the wire tend to move on account of the earth's magnetic field? (In solving think only of the horizontal component of the earth's field, since this is the part of the field at right angles to the wire.)

4. A train is moving toward the north. In doing so, the axis of each truck cuts the earth's magnetic field. Do the currents so generated flow in the axis toward the east or toward the west?

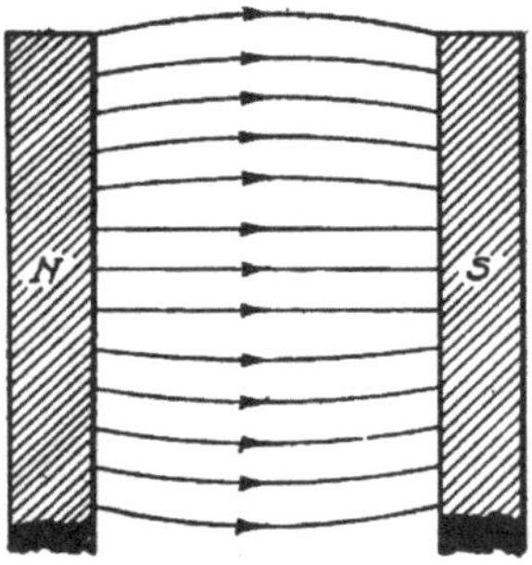

Fig. 147. Field of Magnet Alone

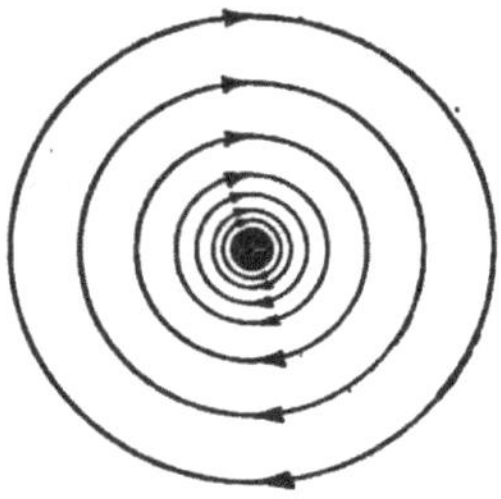

Fig. 148. Field of Current Alone

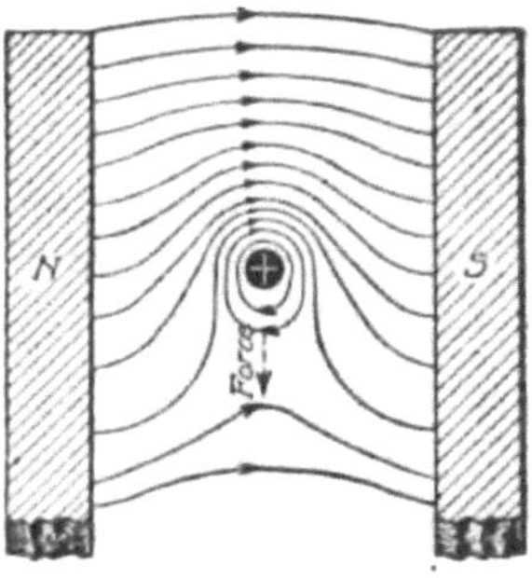

Fig. 149. Combined Field of Magnet and Current

(In solving think only of the vertical component of the earth's field, since this is the part of the field at right angles to the direction of motion of the axis.)

5. If the coil of a sensitive galvanometer is set to swinging while the circuit through the coil is open, it will continue to swing for a long time; but if the coil is shortcircuited, it will come to rest after a very few oscillations. Why? (The experiment may easily be tried. Remember that an e.m.f. is induced in the moving coil. Apply Lenz's law.)

6. A coil of 500 turns of No. 22 copper wire, having a diameter of 30 cm., is held in a vertical plane. If the terminals of the coil are connected to a galvanometer, and the coil is then rotated about a vertical axis, will the current in the coil alternate, as shown by the

galvanometer, when the plane of the coil is east and west, or when it is north and south? (The coil in this problem is cutting the horizontal component of the earth's field.)

7. A coil is thrust over the *N* pole of a magnet. Is the direction of the induced current clockwise or counter-clockwise as you look down upon the pole?

8. A ship having an iron mast is sailing west. Will the top of the mast be + or − due to the induced e.m.f.? If a wire is brought from the top of the mast to the bottom, no current will flow through the circuit. Why?

9. When a wire is cutting lines of force at the rate of 100,000,000 per second, there is induced in it an e.m.f. of 1 volt. A certain dynamo armature has 50 coils of 5 loops each and makes 600 revolutions per minute. Each wire cuts 2,000,000 lines of force twice in a revolution. What is the e.m.f. developed?

INDUCTION COIL AND ITS USES

135. Currents Induced by Currents. In Section 127, currents were induced by the relative motion of a permanent magnet and a coil *C*, Fig. 139. Since the electromagnet *A* of Fig. 150 can be made many times as strong as the permanent magnet of Fig. 139, the induced e.m.f.'s and consequently the induced currents in the coil *C* of Fig. 150 can be made correspondingly greater. According to Lenz's law, when the coil *A* is lowered into the coil *C*, the induced current must set up a field which opposes the motion and hence the current in *C* under these conditions must flow in a direction *opposite* to the current in *A*. For the same reasons, when *A* is withdrawn from *C*, the current in *C* will flow in the *same* direction as that in *A*. The same effects may be produced by making and breaking the circuit at the switch *S*, while the coil *A* is inside the coil *C;* for making the circuit in *A* creates exactly the same magnetic flux through *C* as does thrusting *A* into *C*, and breaking the current in *A* destroys exactly the same magnetic flux through *C*, as does removing *A* from *C*. Thus if *A* is left inside of *C* and an intermittent current is sent through *A* by continuously making and breaking the current at *S*, a current will be induced in *C* which changes direction at each make and break of the current in *A*. The current so induced in *C* is an *alternating* current. The coil *A*, in which the *inducing* current flows, is called the *primary;* and the coil *C*, in which the *induced* current flows, is called the *secondary.*

If a rheostat (variable resistance) be placed in the primary circuit, and the resistance of the circuit be made less, more current will flow through *A*, thus putting more magnetic flux through *C* and inducing a current in *C opposite* to that in *A*. If the resistance of the rheostat in the circuit be made greater, the current in *A* will be made less, thus decreasing the magnetic flux through *C* and inducing in *C* a current in the *same* direction as that in *A*. If the resistance of the rheostat in the circuit be left unchanged, the magnetic flux through *C* will be unchanged and no current will be induced

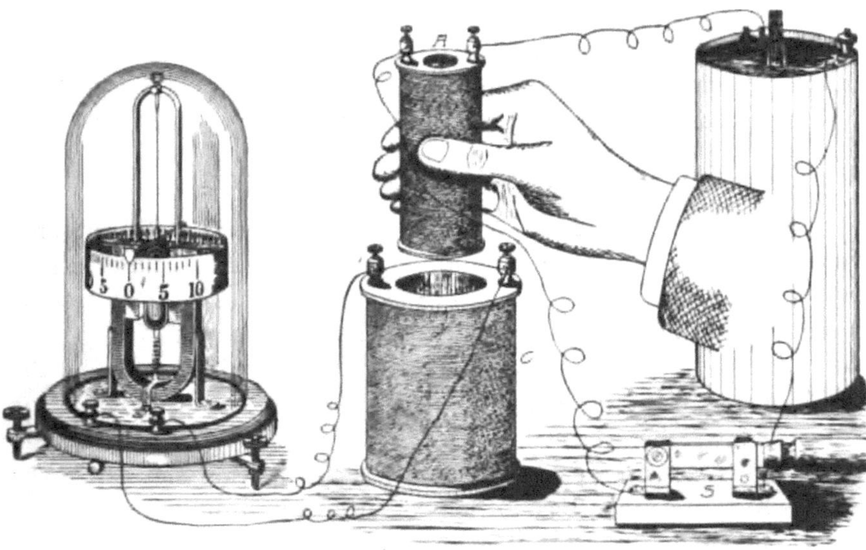

Fig. 150. Experiment Illustrating Currents Induced by Currents

in *C*. Thus we see that the induced currents in *C* depend entirely upon the changes in magnetic flux through *C*, which in turn depends upon the changes of current in *A*. Steady currents (d.c. or direct currents), therefore, cannot be employed when it is desired to transform current from a primary to a secondary. Intermittent currents may be used on the primary, and for this purpose steady currents are made intermittent by various forms of interrupters. The alternating current (a.c.) is the most desirable form of current to use on a primary for many purposes. The a.c. does not necessitate the use of any sort of interrupter, since the current is constantly changing, the flow of the current in one direction putting the desired magnetic flux through the secondary, thus inducing a current in it, and the flow in the opposite direction immediately following creating a magnetic flux through the secondary in the opposite direction, thus inducing a current in the secondary in the opposite direction. Hence, we have an a.c. produced in the secondary by an a.c. in the primary.

If, while coil A is inside of C and a steady current is flowing in A, we insert a core of soft iron into A, we shall make A into a very strong electromagnet, thus greatly increasing the magnetic flux through C and inducing in C a current which *opposes* these magnetic changes, that is, a current in C *opposite* to that in A. Upon removal of the soft iron core, the induced current in C, for the same reason, will flow in the *same* direction as that in A. Thus we see that the currents induced in C are made greater by using stronger currents in A and by using a soft iron core in A.

In all cases, *the induced current in a coil is due to the change in the number of lines of magnetic force (flux) through the coil.*

On making the current in the primary, the current induced in the secondary is opposite in direction to that in the primary.

On breaking the current in the primary, the current induced in the secondary is in the same direction as that in the primary.

136. E.M.F. of the Secondary. If half of the turns of the secondary C, Fig. 150, are unwrapped, the deflection will be found to be just half as great as before. Since the resistance of the circuit has not been changed, we learn from this that *the e.m.f. of the secondary is proportional to the number of turns of wire upon it*—a result which follows also from Section 130. If, then, we wish to develop a very high e.m.f. in the secondary, we have only to make it of a very large number of turns of fine wire. The wire must not, however, be wrapped so far from the core as to include the lines of force which are returning through the air, for when this happens, the coils are threaded in both directions by the same lines, and hence have no currents induced in them.

137. E.M.F. at Make and Break. If the secondary coil C, Fig. 150, be made of 5,000 to 10,000 turns of fine wire, No. 36 or No. 40 copper, and its terminals connected to two small metal cylinders and held in the moistened hands, when the switch S is closed, no shock whatever will be felt, but when the switch is opened, a very strong one will be felt. The experiment shows that the e.m.f. developed at the break of the circuit is enormously greater than that at the make. The explanation is found in the fact that the e.m.f. developed in a coil depends upon the *rate* at which the number of lines of force passing through it is made to change, Section 130. At *make* of the circuit of the primary, the current required an appreciable time, perhaps a tenth of a second, to rise to its full value, just as a current of water, started through a hose, requires an appreciable time to rise to its full height on account of the inertia of the water. An electrical current possesses a property similar to inertia. Hence, the magnetic field about the primary also rises equally gradually to its full strength, and therefore its lines pass into the coil comparatively slowly. At *break*, however, by separating the contact points very quickly, we can make the current in the primary fall to zero in an exceedingly short time, perhaps not more than .00001 second; i.e., we can make all of its lines pass out of the coil in this time. Hence, the *rate* at which lines thread through or cut the secondary is perhaps 10,000 times as great at break as at make, and therefore the e.m.f. is also something like 10,000 times as great. It should be remembered, however, that in a closed secondary, the make current lasts as much longer than the break as its e.m.f. is smaller; hence the total energy of the two is the same, as was indeed indicated by the equal deflections in Section 135, when the current in the primary was made and when it was broken.

138. Self-Induction. We saw in Section 137 that the induced e.m.f. in the secondary at the break of the primary circuit was many times as large as at the make of the primary circuit. Indeed, if the coil *C* had been made a part of the same circuit as the coil *A*, the e.m.f.'s induced in *C* by the changes in the magnetism of the core would of course have been just the same as above. In other words, when a current starts in a coil, the magnetic field which it itself produces tends to induce a current opposite in direction to that of the starting current, that is, tends to oppose the starting of the current; and when a current in a coil stops, the collapse of its own magnetic field tends to induce a current in the same direction as that of the stopping current, that is, tends to oppose the stopping of the current. This means merely that *a current in a coil acts as though it had inertia, and opposes any attempt to start or to stop it.* This inertia-like effect of a coil upon itself is called *self-induction.*

One of the most instructive and striking proofs of the e.m.f. of self-induction is the following experiment: Connect a 110-volt lamp *L*, a large electromagnet *M*, a battery *B* of from 25 to 50 volts, and a key *K*, Fig. 151. When the key *K* is closed, the current divides at *A*, part flowing through the lamp and part energizing the magnet *M*, and unites at *D* returning to the battery. Under these conditions, the glow of the lamp will be very little if any. When *K* is opened, however, the e.m.f. of self-induction will send a large momentary current through the lamp causing it to glow brightly. The arrow at the left of the lamp, Fig. 151, shows the direction of the current from the battery through the lamp when the key is closed. The arrow at the right of the lamp shows the direction of the current through the lamp which is produced by the e.m.f. of self-induction of the magnet coils when the key is opened. The e.m.f. of self-induction produces a spark at the contact points of *K*, and if *K* is opened slowly, this spark will be gradually drawn out producing an arc of resistance sufficient to make the current due to the e.m.f. of self-induction small. It is also a small e.m.f. of self-induction which is produced under these conditions, since the time required for the collapse of the magnetic field has been greatly reduced thus lowering the rate of cutting of the magnet coils by the lines of force. On the other hand, increasing the suddenness of the break at *K* decreases the time required for the magnetic field to collapse, and hence increases the rate of cutting of lines of force by the coil. This may be accomplished by breaking the circuit at *K* with a very quick motion, in which case the lamp will be seen to glow with considerably greater brilliancy.

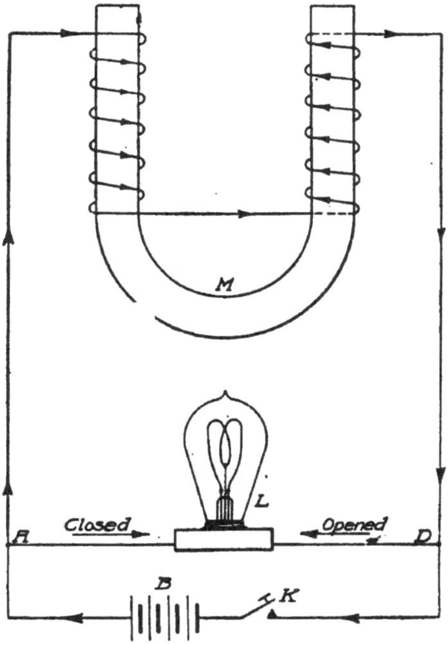

Fig. 151. Diagram Showing the E.M.F. of Self-Induction

139. Self-Induction Applied to Make-and-Break Ignition. While the magneto, Section 175, has been the time-honored device for gas ignition in engines, yet there is another device much cheaper which is just as satisfactory where reliable potentials for operating it are available. This consists of a coil *M*, Fig. 152, of many turns of wire wound on an iron core in order to have a large self-induction and therefore produce a spark when the current from a battery *B*, passing through it is broken by the rotating contact *C*. This produces the spark at *S* when the gas and air mixture is under compression, the explosion driving the piston downward.

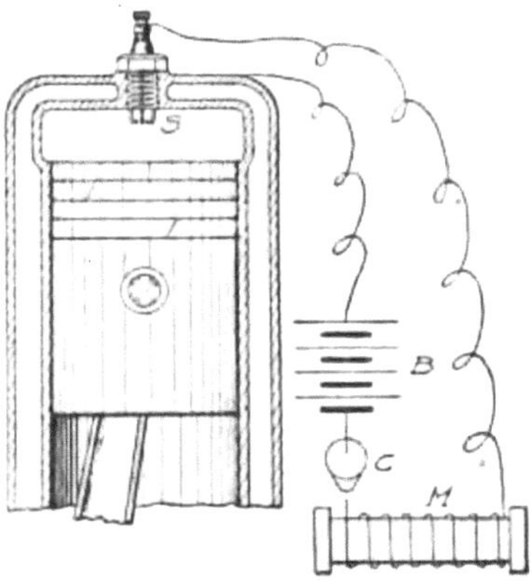

Fig 152 Diagram Showing Make-and-Break Ignition in Gas Engines

The Remy jump-spark coil is similar in operation but is really an induction coil without an interrupter, Section 140, the current being broken by a device like that used above. Gas lights are often arranged so that when a chain is pulled the gas is turned on, and at the same time a current from dry cells is made and broken in a coil which produces a spark between points placed just to the sides and a little above the tip of the gas burner, thus automatically lighting the gas.

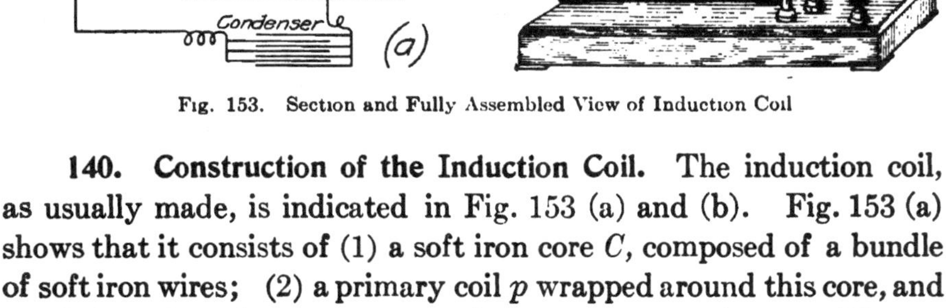

Fig. 153. Section and Fully Assembled View of Induction Coil

140. Construction of the Induction Coil. The induction coil, as usually made, is indicated in Fig. 153 (a) and (b). Fig. 153 (a) shows that it consists of (1) a soft iron core *C*, composed of a bundle of soft iron wires; (2) a primary coil *p* wrapped around this core, and consisting of, perhaps 200 turns of coarse copper wire, e.g., No. 16, which is connected into the circuit of a battery through the contact point at the end of the screw *d;* (3) a secondary coil *s* surrounding the primary in the manner indicated in the diagram, and consisting generally of between 30,000 and 1,000,000 turns of No. 36 copper wire, the terminals of which are the points *t* and *t'*; and (4) a hammer *b*, or other automatic arrangement for making and breaking the circuit of the primary.

Let the hammer *b* be held away from the opposite contact point by means of the finger, then touched to this point, then pulled quickly away. *A spark will be found to pass between t and t' at break only—never at make.* This is because, on account of the opposing influence at make of self-induction in the primary, the magnetic field about the primary rises very gradually to its full strength, and hence its lines pass into the secondary coil comparatively slowly. At *break*, however, by separating the contact points very quickly we can make the current in the primary fall to zero in an exceedingly short time, perhaps not more than .00001 second; that is, we can make all of its lines pass out of the coil in this time. Hence the *rate* at which lines thread through or cut the secondary is perhaps 10,000 times as great at the break as at make, and therefore the e.m.f. is also something like 10,000 times as great. In the normal use of the coil, the circuit of the primary is automatically made and broken at *b* by means of the magnet and the spring *r*, precisely as in the case of the electric bell. Let the student analyze this part of the coil for himself. The condenser, shown in the diagram, with its two sets of plates connected to the conductors on either side of the spark gap between *r* and *d*, is not an essential part of a coil, but when it is introduced, the length of the spark which can be sent across between *t* and *t'* is found to be considerably increased. The reason is as follows When the circuit is broken at *b*, the inertia, or the self-induction, of the primary current, tends to make a spark jump across from *d* to *b;* and if this happens, the current continues to flow through this spark, or arc, until the terminals have become separated through a considerable distance. This makes the current die down gradually instead of suddenly, as it ought to do to produce a high e.m.f. But when a condenser is inserted, as soon as *b* begins to leave *d*, the current begins to flow into the condenser, and this gives the hammer time to get so far away from *d* that an arc cannot be formed. This means a sudden break and a high e.m.f. Since a spark passes between t and t' only at break, it must always pass in the same direction. Coils which give 24-inch sparks (perhaps 500,000 volts) are not uncommon. Such coils usually have hundreds of miles of wire upon their secondaries.

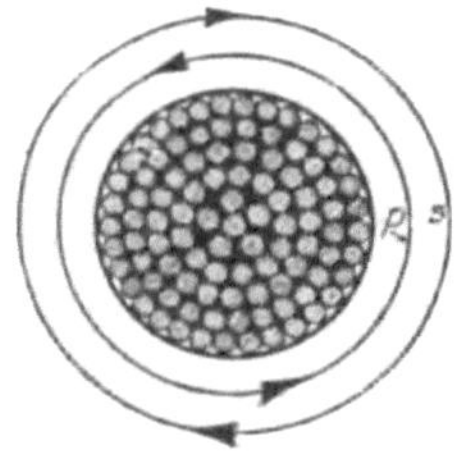
Fig. 154. Section of Laminated Core to Prevent Eddy Currents

141. Laminated Cores. *Foucault Currents.* The core of an induction coil should always be made of a bundle of soft iron wires insulated from one another by means of shellac or varnish, Fig. 154; for whenever a current is started or stopped in the primary p of a coil furnished with a *solid* iron core, Fig. 155, the change in the magnetic field of the primary induces a current in the conducting core C for the same reason that it induces one in the secondary s. This current flows around the body of the core in the same direction as the induced current in the secondary, i.e., in the direction of the arrows. The only effect of these so-called *eddy* or *Foucault* currents is to heat the core. This is obviously a waste of energy. If we can prevent the appearance of these currents, all of the energy which they would waste in heating the core may be made to appear in the current of the secondary. The core is therefore built of varnished iron wires, which run parallel to the axis of the coil, i.e., perpendicular to the direction in which the currents would be induced. The induced e.m.f., therefore, finds no closed circuits in which to set up a current, Fig. 154. It is for the same reason that the iron cores of dynamo and motor armatures, instead of being solid, consist of iron disks placed side by side, as shown in Fig. 156, and insulated from one another by films of oxide. A core of this kind is called a *laminated* core. It will be seen that in all such cores the spaces or slots between the laminae must run at right angles to the direction of the induced e.m.f., i.e., perpendicular to the conductors upon the core.

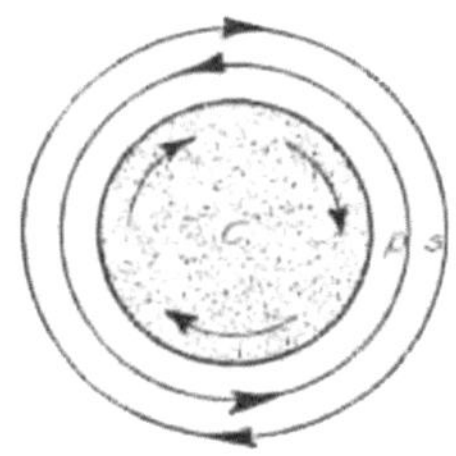
Fig. 155. Diagram Showing Eddy Currents in Solid Core

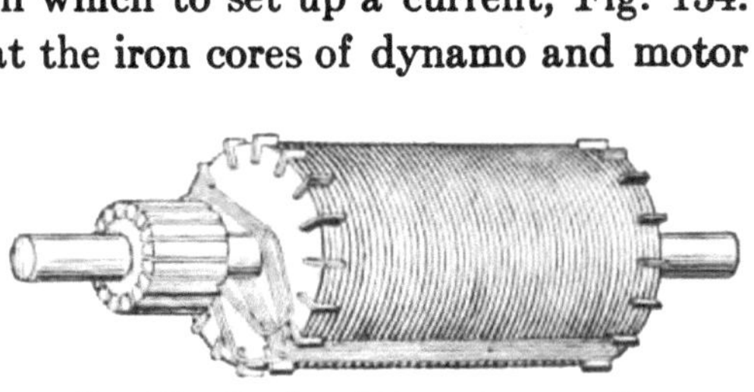
Fig 156. Laminated Drum Armature Core

EXAMPLES FOR PRACTICE

1. Does the spark of an induction coil occur at *make* or at *break?* Why?

2. Explain why an induction coil is able to produce such an enormous e.m.f.

3. Draw a diagram to illustrate the method of operation of the induction coil and fully explain its action.

4. Why should the core of an induction coil be made up of iron wires insulated from each other, rather than of a solid piece of iron?

5. Why will an induction coil work better, giving a higher e.m.f. and a "fatter" spark, when a condenser is connected across the terminals of the interrupter?

INDUCTION-COIL DISCHARGE PHENOMENA

142. Physiological Effects. Physicians use small induction coils for various treatments. If the electrodes of the secondary of such a coil are held in the hands, a muscular contraction is produced, the intensity of which is varied as the patient becomes accustomed to it. Large coils produce a dangerous shock on account of the high e.m.f. which they produce, but when such coils are operated with high-frequency alternating currents, the danger is decreased in proportion to the increase in the number of times per second which the current alternates.

Hence it is that performers on the stage are able to light lamps through their finger tips, draw off long sparks from their noses, etc., these experiments being performed with high-frequency currents of the sort mentioned above.

143. Heating Effects. The heating effects of the discharge are employed in the vibrator of automobiles for ignition purposes. If the terminals of the secondary of the induction coil are separated about 2 centimeters and a bunsen burner is brought up to the gap when the gas is turned on, the passage of the spark through the gas will light the burner. If a spoon containing some ether is placed in the spark gap, the ether will be ignited. Induction coils are also used for igniting at a distance the explosives used for sending off projectiles.

144. Mechanical and Chemical Effects. If a piece of cardboard is placed across the path of the spark, the cardboard will be burred on both sides. Thin plates of any insulating material will be pierced by the discharge. If Leyden jars are placed across the secondary terminals of a very high voltage transformer, say 60,000 volts, the discharge will often break the glass of the Leyden jars.

A very entertaining experiment may be performed in the following way: Moisten a strip of white blotting paper with a solution of potassium iodide and starch paste and place it on a glass plate. Attach one terminal of the induction coil to the blotting paper and the other end to an insulating handle. The latter is then used when the coil is operating to write or draw with the free terminal of the coil upon the blotting paper. A blue mark will be traced on the blotting paper due to the decomposing action of the spark upon the potassium iodide, the blue resulting from the action of the iodine on the starch.

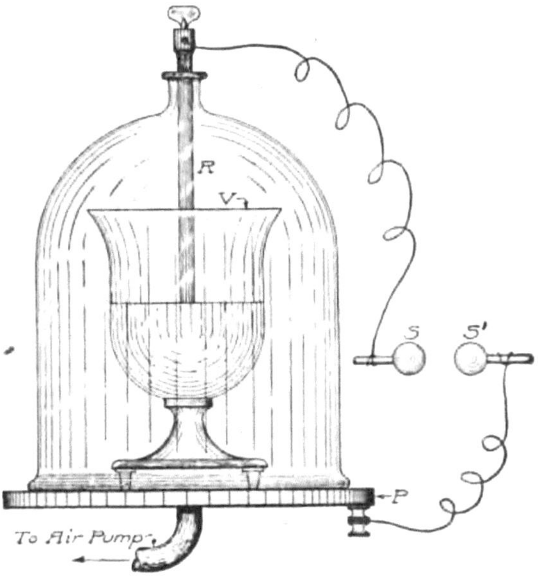

Fig. 157. Gassiot's Cascade Showing Discharge in Partial Vacua

The induction-coil discharge when passed through a sealed tube containing air will produce nitrous acid, by causing the nitrogen and oxygen in the air to unite. This is the principle involved in one of the greatest chemical discoveries of the age, the taking of nitrogen from the inexhaustible supply of air to form ammonia and nitric acid, the basis of innumerable industrial compounds and commercial fertilizers, thus enormously increasing the profits which may be made from our agricultural fields.

145. Discharges in Partial Vacua. *Gassiot's Cascade.* The discharge in a partial vacuum is very strikingly shown by lining a vase *V*, Fig. 157, with tinfoil about halfway to the top. The vase is then placed under the receiver of an air-pump, the receiver being provided with a sliding rod R, or a rod with chain attached to make electrical contact with the tinfoil of the vase. The terminals S and S' of an induction coil are connected to the rod R and the plate P of the air-pump. If S and S' are a few centimeters apart, the discharge will take place between S and S' before any air is exhausted, this being the shorter path over which the discharge will pass. When the air is partially exhausted, however, the discharge will take place between the inside of the vase and the plate P, passing over the top edge of the vase. The latter path is much longer than the path SS' showing that *a discharge takes place through a partial vacuum much more easily than through air at atmospheric pressure.* On account of its resemblance to a cascade and in honor of the originator of the experiment, the phenomenon is called *Gassiot's Cascade.* As the air is more nearly exhausted, the character and beauty of the discharge will vary. If the vase is made of uranium, iron, or lead glass, different color effects will be produced in each case. Other color effects are also produced by admitting other gases into the receiver, such as oxygen, hydrogen, or nitrogen, and again exhausting.

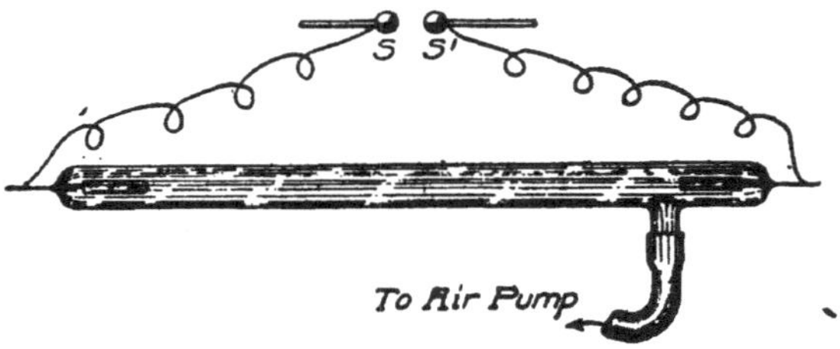

Fig. 158. Simple Geissler Tube for Spark Discharge

The preceding experiment may be shown in principle with the simple tube of Fig. 158, not so strikingly, however, as the cascade effect of the discharge is lacking. Such tubes are made by sealing platinum electrodes into the ends of the glass tube. To get the best result of the luminous discharge, the tube should be exhausted to a pressure of about two millimeters.

Geissler Tubes. Geissler tubes are tubes like that of Fig. 158 except that they are twisted into many fantastic shapes, are made of different kinds of glass, and often surrounded with fluorescent liquids which produce beautiful color effects. Uranium glass, for example, produces a yellow phosphorescence, iron glass produces green, and lead glass, blue. The fluorescent liquids most commonly used to surround the tube are eosin, kerosene, aesculin, uranin, and quinine. A set of Geissler tubes is shown in Fig. 159. The stratifications, or *striae,* shown in the second tube of Fig. 159, are seen to be portions of the illuminated gases which are bright, followed by portions which are dark. Their explanation is beyond the scope of this treatise. They have been produced in tubes fifty feet long.

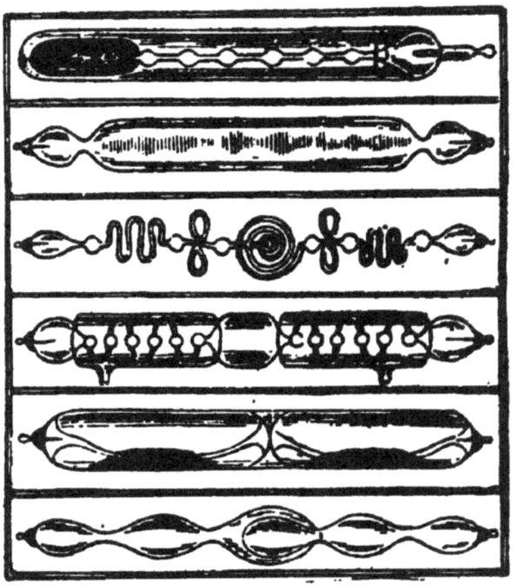

Fig. 159. Typical Geissler Tubes

Discharge of Geissler Tube Is Intermittent. From a sheet of black paper cut disks one-half an inch in diameter. Paste these disks in concentric circles and equally spaced on a larger white disk about ten inches in diameter, Fig. 160, and with the aid of a whirling table or motor rotate the disk in front of a Geissler tube in action in a darkened room. Since the flashes are practically instantaneous, each will give a sharp outline of the position of each spot at that instant. If for any circle of dots the speed of rotation causes the disk to rotate through an angle such that a dot moves through the angle between one or more dots in that circle in the time between flashes, the row of dots will appear to be stationary. If the disk moves through a somewhat larger

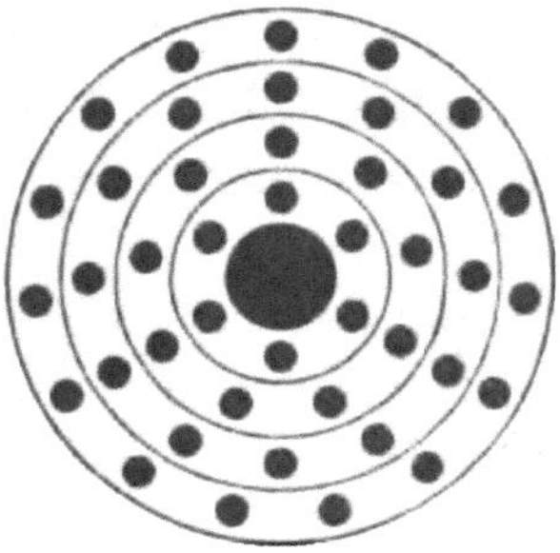

Fig. 160. Disk for Proving Oscillatory Discharge in Geissler Tubes

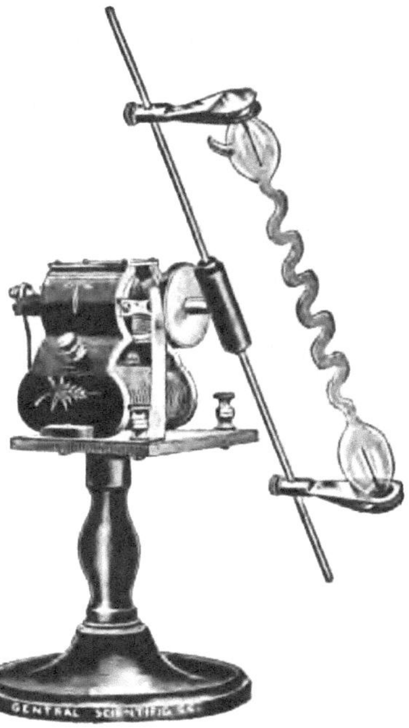

Fig. 161 Motor Driven Apparatus for Displaying Geissler Tubes

angle in the same time, the row of spots in this circle will appear to rotate forward, and if it moves slowly, to rotate backward. Thus, at the proper speed, the spots in one circle will stand still and in the other circles will appear to be rotating forward or else backward.

The especially constructed apparatus shown in Fig. 161 enables one to send a discharge through a Geissler tube while it is being rotated by a motor. The tube will then be seen in a different position for each flash of the light and cross-shaped figures can be produced, the number of images in the cross depending on the speed of rotation of the Geissler tube and the time between flashes in it, the latter depending on the time of vibration of the circuit interrupter.

146. Cathode Rays. When a tube similar to the above is exhausted to a very high degree, say to a pressure of about .001 millimeters of mercury, the character of the discharge changes completely. The glow almost entirely disappears from the residual gas in the tube, and certain invisible radiations called cathode rays begin to be emitted by the cathode (the terminal of the tube which is connected to the negative terminal of the coil or static machine). These rays manifest themselves, first, by the brilliant fluorescent effects which they produce in the glass walls of the tube, or in other substances within the tube upon which they fall; second, by powerful heating effects; and third, by the sharp shadows which they cast.

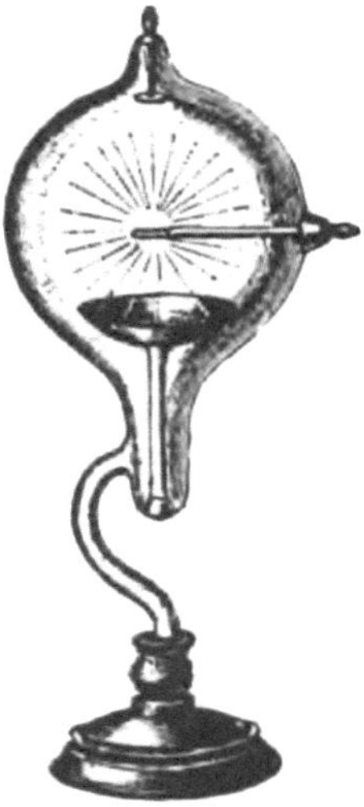

Fig. 162. Heating Effect of Cathode Rays

Thus if the negative electrode is concave, as in Fig. 162, and a piece of platinum foil is placed at the center of the sphere of which the cathode is a section, the rays will come to a focus upon a small part of the foil and will heat it white hot, thus showing that the rays, whatever they are, travel out in straight lines at right angles to the surface of the cathode. This may also be shown nicely by an ordinary bulb of the shape shown in Fig. 164. If the electrode *A* is made the cathode, and *B* the anode, a sharp shadow of the piece of platinum in the middle of the tube will be cast on the wall opposite to *A*, thus showing that the cathode rays, unlike the ordinary electric spark, do not pass between the terminals of the tube, but pass out in a straight line from the cathode surface.

Nature of the Cathode Rays. The nature of the cathode rays was a subject of much dispute between the years 1875, when they first began to be carefully studied, and 1898. Some thought them to be streams of negatively charged particles shot off with great speed from the surface of the cathode, while others thought they were waves in the ether—some sort of invisible light. The following experiment furnishes very convincing evidence that the first view is correct.

NP, Fig. 163, is an exhausted tube within which has been placed a screen *sf* coated with some substance like zinc sulphide which fluoresces brilliantly when the cathode rays fall upon it; *mn* is a mica strip containing a slit *s*. This mica strip absorbs all the cathode rays which strike it; but those which pass through the slit *s* travel the full length of the tube, and although they are themselves invisible, their path is completely traced out by the fluorescence which they excite upon *sf* as they graze along it. If a magnet *M* is held in the position shown, the cathode rays will be seen to be deflected, and in exactly the direction to be expected if they consisted of negatively charged particles.

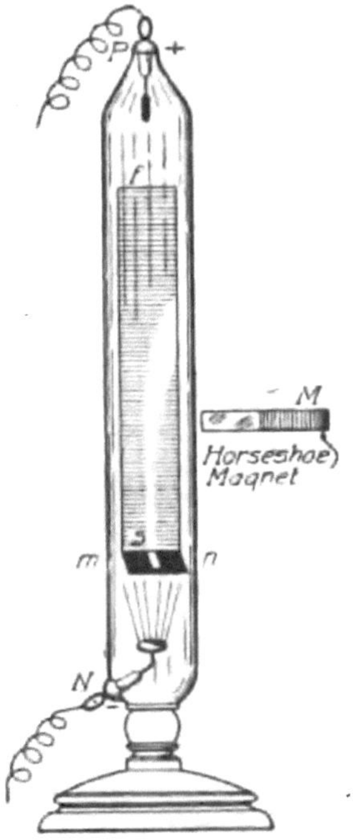

Fig. 163. Deflection of Cathode Rays by a Magnet

When, in 1895, J. J. Thomson of Cambridge, England, proved that the cathode rays were also deflected by electric charges, as was to be expected if they consist of negatively charged particles, and when Perrin in Paris had proved that they actually impart negative charges to bodies on which they fall, all opposition to the projected-particle theory was abandoned.

Size and Velocity of Cathode-Ray Particles. The most remarkable result of the experiments upon cathode rays is the conclusion that the rapidly moving particles of which they consist are not ordinary atoms or molecules, but are, instead, bodies whose mass is only about one two-thousandth the mass of the lightest atom known, namely the atom of hydrogen. Moreover, the velocity with which these particles are projected through the tubes sometimes reaches the stupendous value of 100,000 miles per second, i.e., about one-half the velocity of light.

147. New Theories as to the Constitution of Matter. Furthermore, since experiments of the kind mentioned in Section 146 always lead to the same value for the mass of the cathode-ray particle, no matter what may be the nature of the gas which is used in the bulb, and no matter what may be the nature of the metal which constitutes the cathode, physicists have been forced to the conclusion that these minute particles are constituents of each and every one of the different metallic elements, at least, and probably of all other elements also. It is largely because of these discoveries that the *electron theory* already referred to has been advanced by several of the greatest living physicists. According to J. J. Thomson's formulation of this theory, these cathode particles are the primordial particles out of which the seventy-odd atoms known to chemistry are built up, the chief differences between the various atoms of chemistry consisting in the number of these particles which enter into them. The hydrogen atom, for example, is supposed to contain about 2,000 of these so-called electrons, the oxygen atom 32,000, the mercury atom 400,000, and so on. But since the atoms themselves are probably electrically neutral, it is necessary to assume that they contain equal amounts of positive and negative electricity. Since, however, no evidence has yet appeared to show that positively charged electrons ever become detached from molecules, Thomson brings forward the hypothesis that perhaps the positive charges constitute the nucleus of the atom about the center of which the negative electrons are rapidly rotating.

According to this hypothesis, then, an atom is a sort of infinitesimal solar system whose members, the electrons, are no bigger with respect to the diameter of the atom than is the earth with respect to the diameter of the earth's orbit. Furthermore, according to this hypothesis, it is the vibrations of these electrons which give rise to light and heat waves; it is the streaming through conductors of electrons which have become detached from atoms which constitutes an electric current in a metal; an excess of electrons upon a body constitutes a static negative charge, and a deficiency of electrons constitutes a positive charge, Section 26, *Part* I.

This theory undoubtedly contains many germs of truth. As yet, however, it is in the formative stage and ought to be regarded as a profoundly interesting speculation brought forward by men high in authority in the scientific world, rather than as an established doctrine. However, that such things as negatively charged corpuscles exist, and that they have a mass which is much smaller than that of an atom is now universally admitted.

148. X-Rays. It was in 1895 that the German physicist, Röntgen, first discovered that wherever the cathode rays impinge upon the walls of a tube, or upon any obstacles placed inside the tube, they give rise to another type of invisible radiation which is now known under the name of X-rays, or Röntgen rays. In the ordinary X-ray tube, Fig. 164, a thick piece of platinum P is placed in the center to serve as a target for the cathode rays, which, being emitted at right angles to the con-

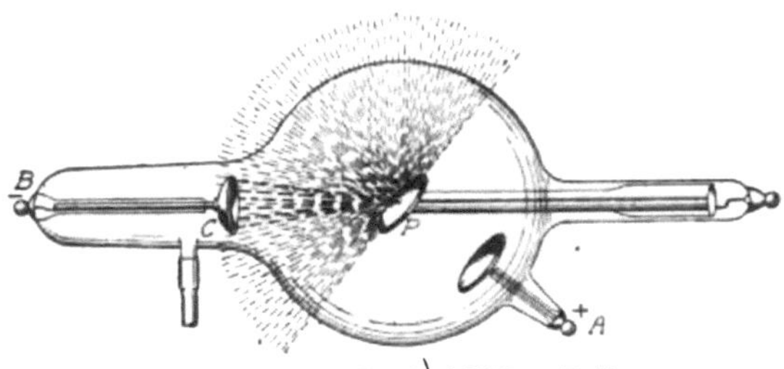

Fig 164. Typical X-Ray Bulb

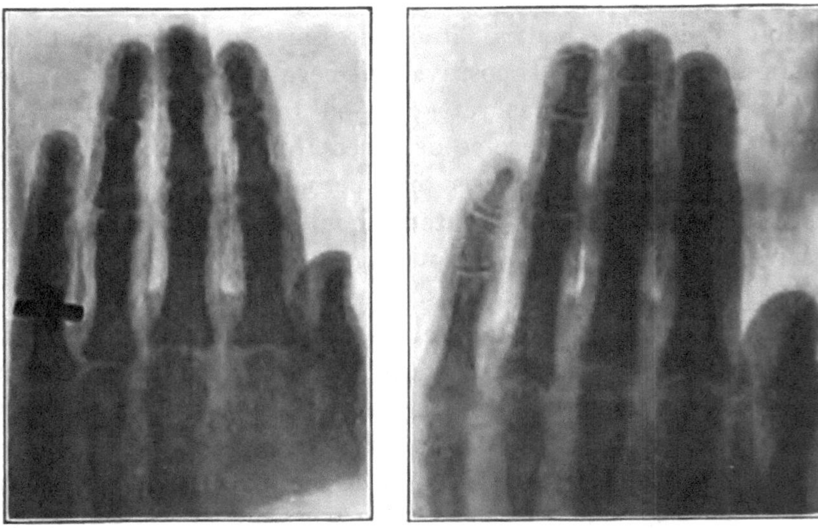

Fig. 165. X-Ray Pictures of Human Hands. Note that Tip of Middle Finger of the Hand at the Right Has Been Cut Off in a Lathe

cave surface of the cathode C, come to a focus at a point on the surface of this plate. This is the point at which the X-rays are generated and from which they radiate in all directions.

In order to convince oneself of the truth of this statement, it is only necessary to observe an X-ray tube in action. It will be seen to be divided into two hemispheres by the plane which contains the platinum plate, Fig. 164. The hemisphere which is facing the source of the X-rays will be aglow with a greenish fluorescent light, while the other hemisphere, being screened from the rays, is darker. The fluorescence is due to an effect which the X-rays have upon the glass. In this respect they are like cathode rays. By moving a fluoroscope (a zinc-sulphide screen) about the tube, it will be made evident that the rays which render the bones visible, Fig. 165, come from *P*.

149. Nature of X-Rays. While X-rays are like cathode rays in producing fluorescence, and also in that neither of them can be reflected, refracted, or polarized, as is light, they nevertheless differ from cathode rays in several important respects. First, X-rays penetrate many substances which are quite impervious to cathode rays; for example, they pass through the walls of the glass tube, while cathode rays ordinarily do not. Again, X-rays are not deflected either by a magnet or by an electrostatic charge, nor do they carry electrical charges of any sort. Hence, it is certain that they do not consist, like cathode rays, of streams of electrically charged particles. Their real nature is still unknown, but they are at present generally regarded as irregular pulses in the ether, set up by the sudden stopping of the cathode-ray particles when they strike an obstruction.

150. X-Rays Render Gases Conducting. One of the notable properties which X-rays possess in common with cathode rays is the property of causing any electrified body on which they fall to lose its charge slowly.

To demonstrate the existence of this property, let any X-ray bulb be set in operation within five or ten feet of a charged gold-leaf electroscope. The leaves at once begin to fall together.

The reason for this is that the X-rays shake loose electrons from the atoms of the gas and thus fill it with positively and negatively charged particles, each negative particle being at the instant of separation an electron and each positive particle an atom from which an electron has been detached. Any charged body in the gas, therefore, draws toward itself charges of sign opposite to its own, and thus becomes discharged.

151. X-Ray Pictures. The most striking property of X-rays is their ability to pass through many substances which are wholly opaque to light, such, for example, as cardboard, wood, leather, flesh, etc. Thus, if the hand is held close to a photographic plate and then exposed to X-rays, a shadow picture of the denser portions of the hand, i.e., the bones, is formed upon the plate. In Fig. 165 are shown copies of such pictures.

RADIOACTIVITY

152. Discovery of Radioactivity. In 1896, Henri Becquerel performed the following experiment in Paris. He wrapped up a photographic plate in a piece of perfectly opaque black paper, laid a coin on top of the paper, and suspended above the coin a small quantity of the mineral uranium. He then set the whole away in a dark room and let it stand for several days. When he developed the photographic plate, he found upon it a shadow picture of the coin similar to an X-ray picture. He concluded, therefore, that *uranium possesses the property of spontaneously emitting rays of some sort which have the power of penetrating opaque objects and of affecting photographic plates just as X-rays do.* He also found that these rays, which he called *uranium rays,* are like X-rays in that they discharge electrically charged bodies on which they fall. He discovered, too, that the rays are emitted by all uranium compounds.

153. Radium. It was but a few months after Becquerel's discovery that Madame Curie began an investigation in Paris of all the known elements, to find whether any of the rest of them possessed the remarkable property which had been found to be possessed by uranium. She found that one of the remaining known elements, namely thorium, the chief constituent of Welsbach mantles, is capable, together with its compounds, of producing the same effect. After this discovery, the rays from all this class of substances began to be called *Becquerel rays,* and all substances which emitted such rays were called *radioactive* substances.

But, in connection with this investigation, Madame Curie noticed that pitchblende, the crude ore from which uranium is extracted, consisting largely of uranium oxide, would discharge her electroscope about four times as fast as pure uranium. She inferred, therefore, that the radioactivity of pitchblende could not be due solely to the uranium contained in it, and that pitchblende must contain some hitherto unknown element which has the property of emitting Becquerel rays more powerfully than uranium or thorium. After a long and difficult search, she succeeded in separating from several tons of pitchblende a few hundreths of a gram of a new element which was capable of discharging an electroscope more than a million times as rapidly as either uranium or thorium. She named this new element *radium.*

154. Nature of Becquerel Rays. That these rays which are spontaneously emitted by radioactive substances are not X-rays, in spite of their similarity in affecting a photographic plate, in causing fluorescence, and in discharging electrified bodies, is proved by the fact that they are found to be deflected by both magnetic and electric fields, and further by the fact that they impart electric charges to bodies upon which they fall. These properties constitute strong evidence that *radioactive substances project from themselves electrically charged particles.*

Experiments by Rutherford and others have established that these rays consist of three different types, with very distinctive properties. A further study of them is not deemed wise; however, the fundamental idea of the actual projection, from a mass of radium, of material particles of almost infinitesimal size at enormous velocities can easily be grasped, and serves to give the student an idea of how the modern scientist is broadening our knowledge of those phenomena which used to be called the "mysteries" of Nature.

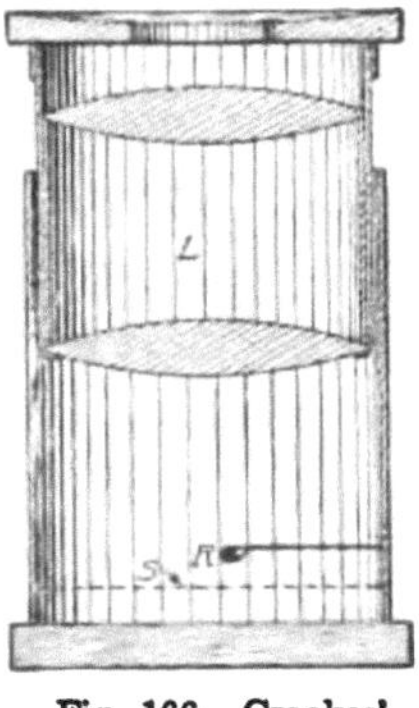

Fig. 166. Crookes' Spinthariscope

155. Crookes' Spinthariscope. In 1903, Sir William Crookes devised a little instrument called the *spinthariscope,* which furnishes very direct and striking evidence that particles are being continuously shot off from radium with enormous velocities. Radium itself glows in the dark with a light which resembles that of the glowworm, and when placed near certain substances like zinc sulphide, it causes them to light up with a glow which is more or less brilliant, according to the amount of radium at hand. In the spinthariscope a tiny speck of radium *R*, Fig. 166, is placed about a millimeter above a zinc-sulphide screen *S*, and the latter is then viewed through a lens *L*, which gives a magnification of from ten to twenty diameters. The continuous soft glow of the screen, which is all one sees with the naked eye, is resolved by the microscope into a thousand tiny flashes of light. The appearance is as though the screen were being fiercely bombarded by an incessant rain of projectiles, each impact being marked by a flash of light, just as sparks fly from a flint when struck with steel. The experiment is a very beautiful one, and it furnishes very direct and convincing evidence that radium is continually projecting particles from itself at stupendous speed.

156. The Disintegration of Radioactive Substances. Whatever may be the cause of this ceaseless emission of particles exhibited by radioactive substances, it is certainly not due to any ordinary chemical reactions; for Madame Curie showed, when she discovered the activity of thorium, that the activity of all the radioactive substances is simply proportional to the amount of the active element present, and has nothing whatever to do with the nature of the chemical compound in which the element is found. Thus, thorium may be changed from a nitrate to a chloride or a sulphide, or it may undergo any sort of chemical reaction, without any change whatever being noticeable in its activity. Furthermore, radioactivity has been found to be independent of all *physical* as well as *chemical* conditions. The lowest cold or the greatest heat does not appear to affect it in the least. Radioactivity, therefore, seems to be as unalterable a property of the atoms of radioactive substances as is weight itself. For this reason Rutherford has advanced the theory that the atoms of radioactive substances are slowly disintegrating into simpler atoms. Uranium and thorium have the heaviest atoms of all the elements. For some unknown reason they seem not infrequently to become unstable and project off a part of their mass. What is left of the atom is a new substance with chemical properties different from those of the original atom. This new atom is, in general, also unstable and breaks down into something else. This process is repeated over and over again until some stable form of atom is reached. Somewhere in the course of this atomic catastrophe some electrons leave the mass.

According to this point of view, which is now generally accepted, radium is simply one of the stages in the disintegration of the uranium atom. The atomic weight of uranium is 238.5, that of radium about 226, that of helium 3.994. Radium would then be uranium after it had lost 3 helium atoms. The further disintegration of radium through four additional transformations has been traced. It has been conjectured that the fifth and final one is lead. If we subtract 8×3.994 from 238.5, we obtain 206.5, which is very close to the accepted value for lead, namely 207. In a similar way, six successive stages in the disintegration of the thorium atom (atomic weight 232) have been found, but the final product is unknown.

157. Energy Stored Up in the Atoms of the Elements. In 1903, the two Frenchmen, Curie and Labord, made an epoch-marking discovery. It was that radium is continually evolving heat at the rate of about 100 calories per hour per gram. (More recent measurements have given 118 calories.) This result was to have been anticipated from the fact that the particles which are continually flying off from the disintegrating radium atoms subject the whole mass to an incessant internal bombardment which would be expected to raise its temperature. This measurement of the exact amount of heat evolved per hour enables us to estimate how much heat energy is evolved in the disintegration of one gram of radium. It is about two thousand million calories—fully three hundred thousand times as much as is evolved in the combustion of one gram of coal. Furthermore, it is not impossible that similar enormous quantities of energy are locked up in the atoms of *all* substances, existing there perhaps in the form of the kinetic energy of rotation of the electrons. The most vitally interesting question which the physics of the future has to face is : Is it possible for man to gain control of any such store of subatomic energy and to use it for his own ends? Such a result does not now seem likely or even possible; and yet the transformations which the study of physics has wrought in the world within a hundred years were once just as incredible as this.

RADIOTELEGRAPHY

158. Proof That the Discharge of a Leyden Jar Is Oscillatory. It is a well-known fact that the sound waves sent out by a sounding tuning-fork will set into vibration a similar fork placed a few feet away, provided the latter fork has the same natural period as the former. The following is the complete electrical analogy of this experiment: Connect the inner and the outer coats of a Leyden jar *A*, Fig. 167, by a loop of wire *cdef*, the sliding cross-piece *de* being arranged so that the length of the loop may be altered at will. Also let a strip of tinfoil be brought over the edge of this jar from the inner coat to within about 1 millimeter of the outer coat at *C*. Connect the two coats of an exactly similar jar *B* with the knobs *n* and *n′* by a second similar wire loop of fixed length. Place the two jars side by

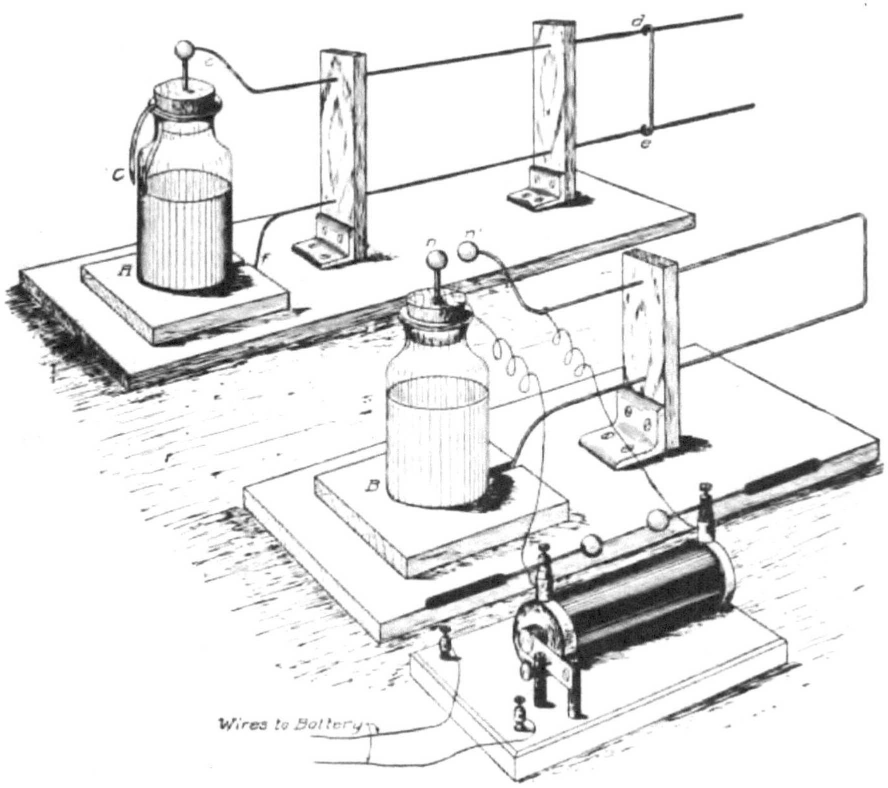

Fig. 167. Illustrating Tuning of Electrical Vibration

side with their loops parallel, and charge and discharge the jar *B* successively, by connecting its coats with a static machine or an induction coil. At each discharge of jar *B* through the knobs *n* and *n′*, a spark will appear in the other jar at *C*, *provided the cross-piece de is so placed that the areas of the two loops are the same.* When *de* is slid along so as to make one loop considerably larger or smaller than the other, the spark at *C* disappears. The natural period of two circuits is the same whenever the product of the capacity and the inductance of one circuit is equal to that of the other.

The experiment, therefore, demonstrates that two electrical circuits, like two tuning-forks, can be *tuned* so as to respond to each other sympathetically, and that just as the tuning-forks will cease to respond as soon as the period of one is slightly altered, so this *electrical resonance* disappears when the exact symmetry of the two circuits is destroyed. Since, obviously, this phenomenon of resonance can occur only between systems which have *natural periods* of vibration, the experiment proves that the discharge of a Leyden jar is a vibratory, i.e., an oscillatory, phenomenon. As a matter of fact, when such a spark is viewed in a rapidly revolving mirror, it is actually found to consist of from ten to thirty flashes following at equal intervals of about .0000001 of a second. Fig. 168 is a photograph of such a spark.

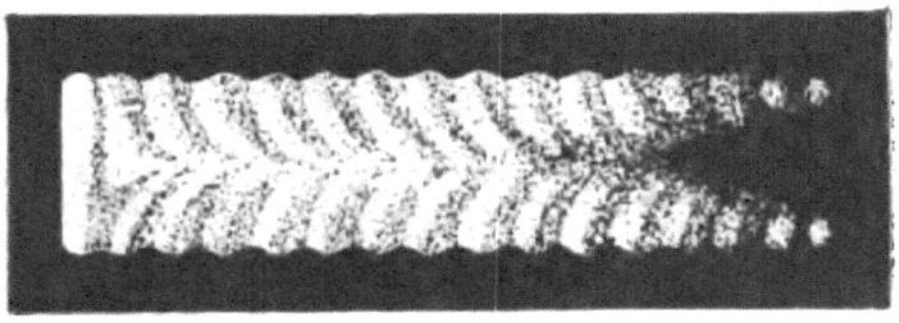

Fig. 168. Oscillations of an Electric Spark

In spite of these oscillations, the whole discharge may be made to take place in the incredibly short time of .0000001 of a second. This fact coupled with the extreme brightness of the spark has made possible the surprising results of so-called *instantaneous electric-spark photography*, as shown by Fig. 169, which is a series of photographs of the passage of a bullet through a soap bubble. The film was rotated continuously instead of intermittently, as in ordinary moving-picture photography. The illuminating flashes, 5,000 per second, were so nearly instantaneous that the outlines are not blurred.

159. Electric Waves. The experiment described in Section 158 demonstrates not only that the discharge of a Leyden jar is oscillatory, but also that these electrical oscillations set up in the surrounding medium are disturbances, or waves of some sort, which travel to a neighboring circuit and act upon it precisely as the air waves acted on the second tuning-fork in the sound experiment. Whether these are waves in the air, like sound waves, or disturbances in the ether, like light waves, can be determined by measuring their velocity of propagation. The first determination of this velocity was made by Heinrich Hertz in Germany in 1888. He found it to be precisely the same as that of light, i.e., 300,000 kilometers per second. *This result shows, therefore, that electrical oscillations set up waves in the ether.* These waves are now known as Hertz waves.

The length of the waves in the experiment described in Section 158 must evidently have been very great, viz, about $\frac{300,000,000}{10,000,000} =$

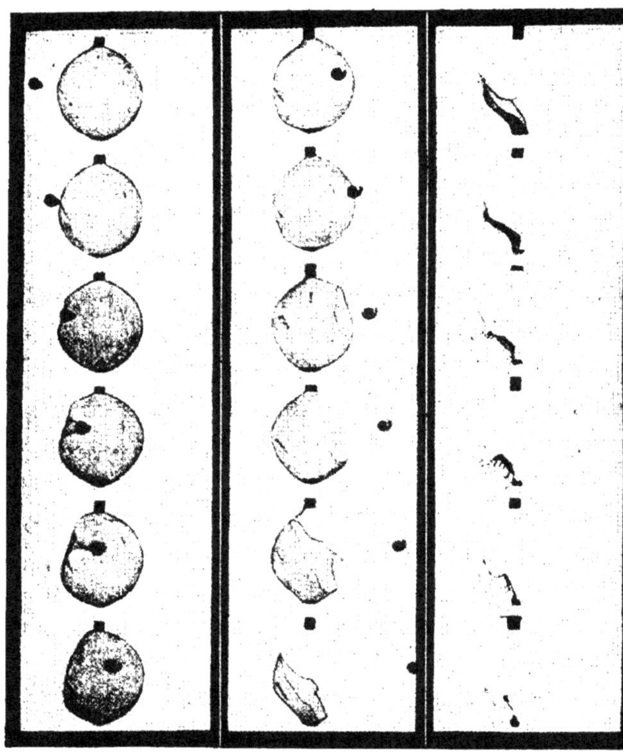

Fig. 169. Moving Picture of Soap Bubble Punctured by a Bullet. Exposures Made under Electric Spark Discharge
From "Moving Pictures" by F. A. Talbot, Courtesy of J. B. Lippincott Company

30 meters, since the velocity of light is 300,000,000 meters per second, and since there are 10,000,000 oscillations per second; for we know that for any wave motion, wave length is equal to velocity divided by the number of oscillations per second. By diminishing the size of the jar and the length of the circuit, the length of the waves may be greatly reduced. By causing the electrical discharges to take place between two balls, only a fraction of a

millimeter in diameter, instead of between the coats of a condenser, electrical waves have been obtained as short as .3 centimeter only fifty times as long as the longest measured heat waves. If the discharge is made to take place between large spheres and a long spark gap, the electrical waves sent off are often many miles in length. It is this latter type of electrical wave which is used in long-distance radiotelegraphy.

160. Coherer. In the experiment described in Section 158, we detected the presence of the electrical waves by means of a small spark gap *C*, in a circuit almost identical with that in which the oscillations were set up. This same means may be employed for the detection of waves many feet away from the source, but the instrument which was early used for this purpose is the *coherer*. Its principle is illustrated in the following experiment:

Let a glass tube several centimeters long and 6 or 8 millimeters in diameter be filled with fine brass or nickel filings, and let copper wires be thrust into these filings to within a distance of about a centimeter of each other. Let these wires be connected in series with a Daniell cell and a simple D'Arsonval galvanometer. The resistance of the loose contacts of the filings will be so great that very little current will flow through the circuit. Now let the static machine or induction coil be started a few feet away. The galvanometer will show a strong deflection as soon as a spark passes between the knobs of the electrical machine. This is because the electric waves, as soon as they fall upon the filings, cause them to cohere or cling together, so that the electrical resistance of the tube of filings is reduced to a small fraction of what it was before. If the tube is tapped with a pencil, the old resistance will be restored, because the filings have been broken apart by the jar. The experiment may then be repeated.

161. Simple Wireless Outfit. The experiment described in Section 160 illustrates completely the method of transmitting messages by wireless telegraphy. The transmitter, or oscillator, consists of an induction coil, Fig. 170, between the knobs of which, *n* and *n′*, the oscillatory electric spark passes as soon as the circuit of the primary is closed by depressing the key *K*. This spark sends out waves into the ether, which are detected at the receiving station, in some cases thousands of miles away, by means of a coherer not differing at all in principle from that described in the last paragraph. This coherer *C*, Fig. 171, is in circuit with a relay *R*. When the electrical waves fall upon the coherer circuit, the resistance of this circuit is greatly reduced, and hence the battery *B* pulls down the armature *A* and thus closes the circuit of the bell, or sounder *D*, through the contact point *P*. Hence the bell starts at once to ring. But

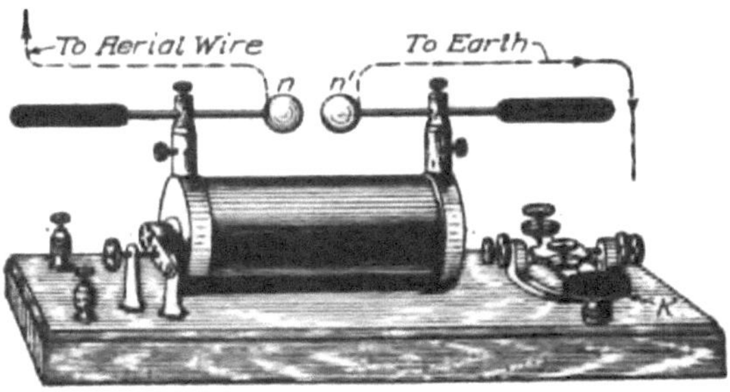

Fig. 170. Wireless Transmitting Apparatus

in this operation of ringing, the clapper *M* strikes against the coherer tube *C* and thus causes the filings to *de-cohere*. The spring *S* then draws back the armature *A*, and the instrument is in condition to receive another signal. Thus, for every spark which is sent out by the coil at the sending station there is a click of the bell, or sounder, at the receiving station. It is found that the efficiency of both transmitter and receiver is very greatly increased by grounding one terminal of each, and connecting to the other terminal a wire which runs up vertically into the air (dotted lines, Figs. 170 and 171). This wire is sometimes between one hundred and two hundred feet high. For sending signals across a schoolroom, wires ten or twelve feet high will be found an advantage, though they are not essential. Silver or nickel filings will be found to work best in the coherer. One Leclanché or dry cell is sufficient for *B* and *B'*.

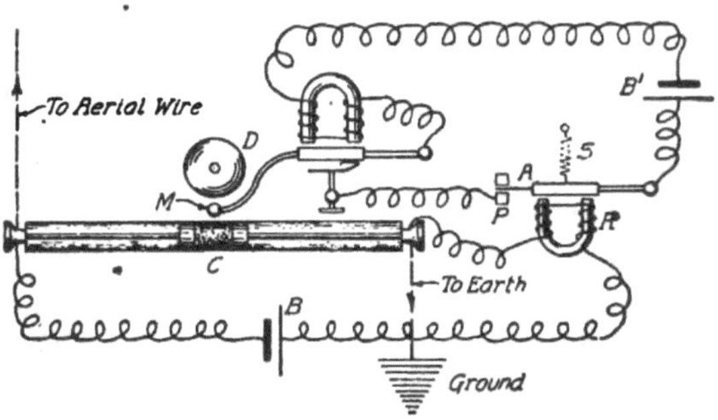

Fig. 171. Wireless Receiving Apparatus

162. Modern Wireless Outfit.* The last experiment illustrates completely the method of transmitting wireless messages during the first decade after Marconi, in 1896, had realized commercial wireless telegraphy. At present the essential elements of the Marconi system of wireless telegraphy are as follows:

*For one of several practical treatises on Wireless Telegraphy, the reader is referred to Experimental Wireless Stations", by Philip E. Edelmann, Minneapolis, Minnesota.

The transmitter, Fig. 172, consists of an ordinary induction coil or transformer T_1, through the primary of which a current is sent from the alternator A. The secondary S of this transformer charges the condenser C_1 until its potential rises high enough to cause a spark discharge to take place across the gap s. This discharge of C_1 is oscillatory, Section 158, the frequency being of the order of 1,000,000 per second, but subject to the control of the operator through the sliding contacts c, precisely as in the case illustrated in Fig. 167. The oscillations in this condenser circuit induce oscilla-

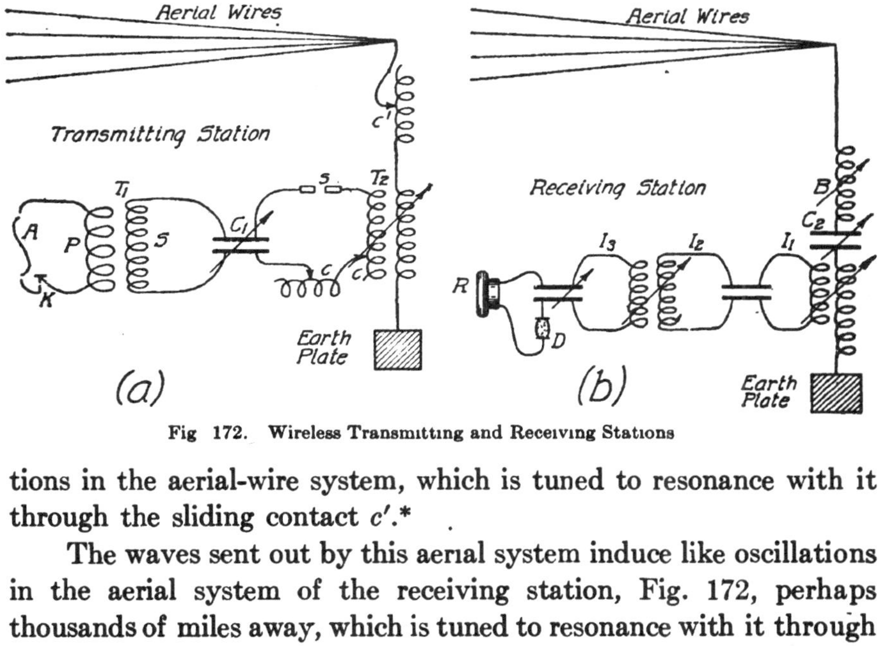

Fig 172. Wireless Transmitting and Receiving Stations

tions in the aerial-wire system, which is tuned to resonance with it through the sliding contact c'.*

The waves sent out by this aerial system induce like oscillations in the aerial system of the receiving station, Fig. 172, perhaps thousands of miles away, which is tuned to resonance with it through the variable capacity C_2 and "inductance" B. These oscillations induce exactly similar ones in the condenser circuits I_1, I_2, and I_3, all of which are tuned to resonance with the receiving aerial system. The detector of the oscillations in I_3 is simply a crystal of carborundum D in series with a telephone receiver R. This crystal, like the mercury arc rectifier has the property of transmitting a current in one direction only. Were it not for this property, the telephone

*In the diagram an arrow drawn diagonally across a condenser indicates that for the sake of tuning, the condenser is made adjustable. Similarly, an arrow across two circuits coupled inductively, like the primary and secondary of the "oscillation transformer" T_2, indicates that the amount of interaction of the two circuits can be varied, as, for example, by sliding one coil a larger or a smaller distance inside the other.

could not be used as a detector because the frequency is so high—of the order of a million. In view of this property, however, while the oscillations of a given spark last, an intermittent current passes in one direction and then ceases until the oscillations of the next spark arrive. Since from 300 to 1,000 sparks pass at *s* per second when the key K is closed, the operator hears a musical note of this frequency as long as K is depressed. Long and short notes then correspond to the dots and dashes of ordinary telegraphy.

The stretching of the aerial wires horizontally instead of vertically, as was formerly done, permits to some extent of directive sending and receiving, for as in the experiment of Section 158, the sending and receiving wires work best when they are parallel.

The three tuned circuits, I_1, I_2, I_3, are used because such a series of tuned circuits does not pick up waves of other periods. For "nonselective" receiving these circuits are omitted and the detector and telephone are placed directly across the condenser C_2. The resistance of the telephone is so high that it does not interfere with the oscillations of the condenser system across which it is placed.

163. Electromagnetic Theory of Light. The study of electromagnetic radiations like those discussed in the preceding paragraphs, has shown not only that they have the speed of light, but that they are reflected, refracted, and polarized; in fact, that they possess all the properties of light waves, the only apparent difference being in their greater wave length. Hence *modern physics regards light as an electromagnetic phenomenon;* that is, light waves are thought to be generated by the oscillations of the electrically charged parts of the atoms. It was as long ago as 1864 that Clerk Maxwell, of Cambridge, England, one of the world's most brilliant physicists and mathematicians, showed that it ought to be possible to create ether waves by means of electrical disturbances. But the experimental confirmation of his theory did not come until the time of Hertz's experiments in 1888. Maxwell and Hertz together, therefore, share the honor of establishing the modern electromagnetic theory of light.

TELEPHONE

164. Simple Telephone without Batteries. The telephone was invented in 1875 by Elisha Gray, of Chicago, and Alexander Graham Bell, of Washington. In its simplest form it consists, at each end, of a permanent bar magnet A, Fig. 173, surrounded by a coil of fine wire B, in series with the line, and an iron disk, or diaphragm E mounted close to one end of the magnet. When a sound is made in front of the diaphragm, the vibrations produced by the sounding body are transmitted by the air to the diaphragm, thus causing the latter to vibrate back and forth in front of the magnet.

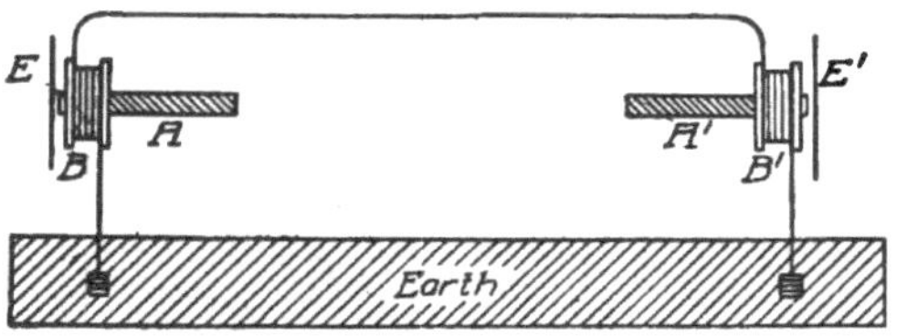

Fig. 173. Simple Telephone

These vibrations of the diaphragm produce slight backward and forward movements of the lines of force which pass into the disk from the magnet in the manner shown in Fig. 174. Some of these lines of force, therefore, cut across the coil B, Fig. 173, first in one direction and then in the other, and in so doing induce currents in it. These induced currents are transmitted by the line to the receiving station, where those in one direction pass around B' in such a way as to *increase* the strength of the magnet A', and thus increase the pull which it exerts upon E'; while the opposite currents pass around B' in the opposite direction, and therefore *weaken* the magnet A' and diminish its pull upon E'. When, therefore, E moves in one direction, E' also moves in one direction, and when E reverses its motion, the direction of motion of E' is also reversed. In other words, the induced currents, transmitted by the line cause E' to reproduce the motions of E. E' therefore sends out sound waves exactly like those which fell upon E. In exactly the same way, a sound made in front of E' is reproduced at E. Telephones of this simple type will work satisfactorily for a distance of several miles. This simple form of instrument is still used at the receiving end of the modern telephone, the

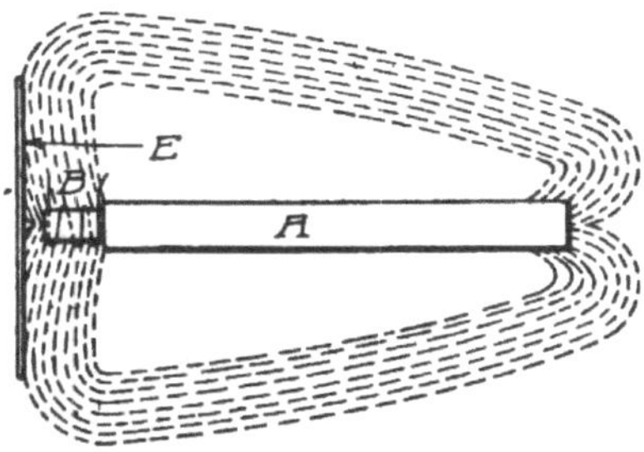

Fig. 174. Magnetic Field in a Telephone Receiver

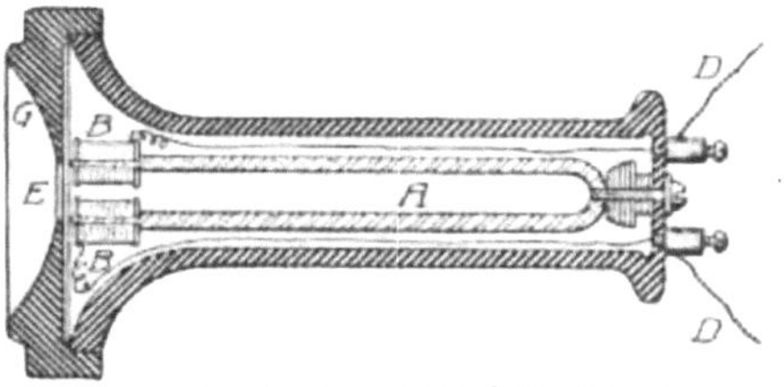

Fig. 175. Section of Modern Telephone Receiver

only improvement which has been introduced consisting in the substitution of a U-shaped magnet for the bar magnet. The instrument used at the transmitting end has, however, been changed, as will be explained in the next paragraph, and the circuit is now completed

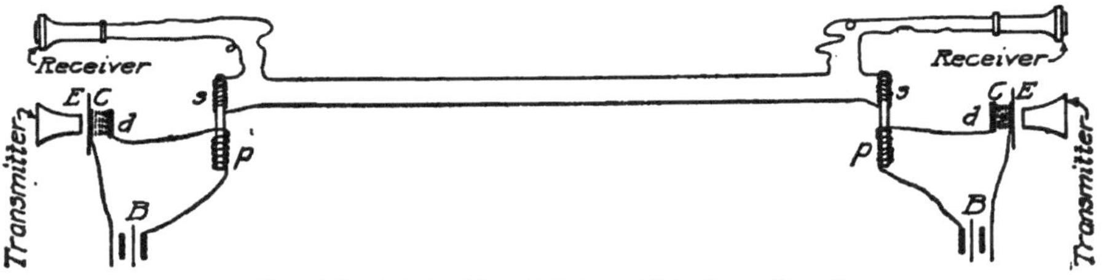

Fig. 176. Typical Local Battery Telephone Circuit

through a return wire instead of through the earth. A modern telephone receiver is shown in Fig. 175; *G* is the earpiece, *E* the diaphragm, *A* the U-shaped magnet, *B* the coils (consisting of many turns of fine wire), and *D* the terminals of the line.

165. Modern Transmitter. To increase the distance at which telephoning may be done, it is necessary to increase the strength of the induced currents. This is done in the modern transmitter by replacing the magnet and coil by an arrangement which is essentially an induction coil, the current in the primary of which is caused to vary by the motion of the diaphragm. This is accomplished as follows:

The current from the battery *B*, Fig. 176, is led first to the back of the diaphragm *E*, whence it passes through a little chamber *C* filled with granular carbon to the conducting back *d* of the transmitter, and thence through the primary *p* of the induction coil, and back to the battery. As the diaphragm vibrates, it varies the pressure upon the many contact points of the granular carbon through which the primary current flows. This produces considerable variation in the resistance of the primary circuit, so that as the diaphragm moves forward, i.e., toward the carbon, a comparatively large current flows through *p*, and as it moves back a much smaller current. These changes in the current strength in the primary *p* produce changes in the magnetism of the soft-iron core of the induction coil.

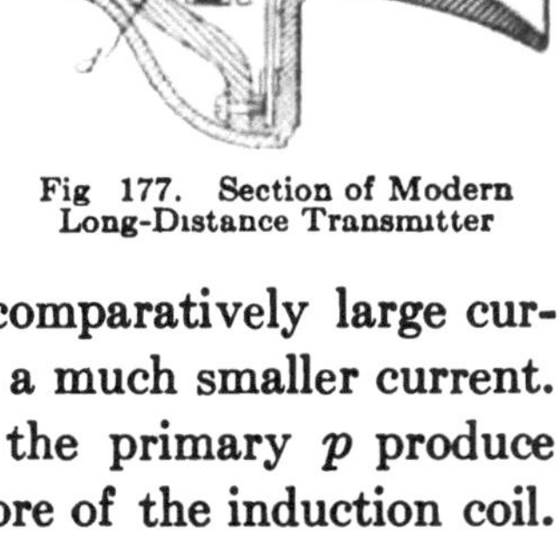

Fig 177. Section of Modern Long-Distance Transmitter

Currents are therefore induced in the secondary *s* of the induction coil, and these currents pass over the line and affect the receiver at the other end in the manner explained in Section 40. Fig. 177 shows the cross-section of a complete long-distance transmitter.

166. Subscriber's Telephone Connections. In the most recent practice of the Bell Telephone Company, the local battery at the subscriber's end is done away with altogether and the primary current is furnished by a 24-volt battery at the central station. Fig. 178 shows the essential elements of such a system. When the subscriber wishes to call up central, he has only to lift the receiver from the hook. This closes the line circuit at *t*, and the direct current which at once begins to flow from the battery *B* through the electromagnet *g*, closes the circuit of *B* through the glow lamp *l*, and the contact point *r*. This lights up the lamp *l* which

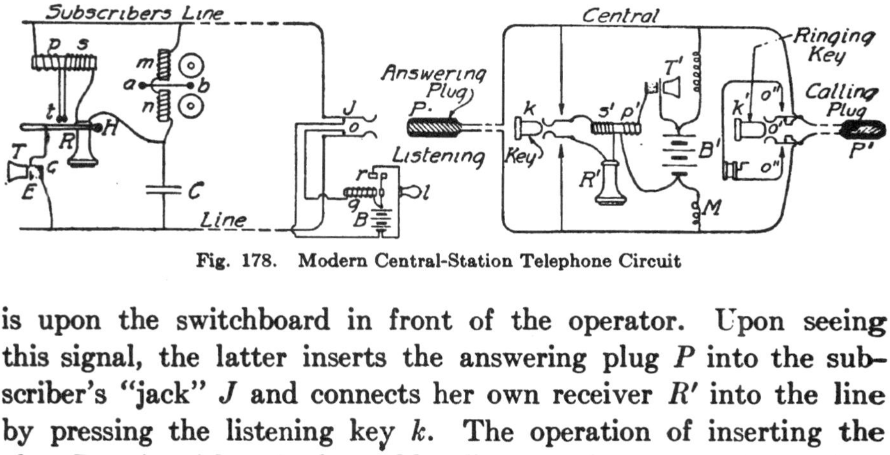

Fig. 178. Modern Central-Station Telephone Circuit

is upon the switchboard in front of the operator. Upon seeing this signal, the latter inserts the answering plug *P* into the subscriber's "jack" *J* and connects her own receiver R' into the line by pressing the listening key *k*. The operation of inserting the plug *P* extinguishes the lamp *l* by disconnecting the contact points *o*. The battery B' is, however, now upon the line (*B* and B' are, in fact, one and the same battery, shown here as separate, only for the sake of simplifying the diagram). As, now, the subscriber talks into the transmitter *T*, the strength of the direct current from the battery B' through the primary *p* is varied by the varying pressure of the diaphragm *E* upon the granular carbon *c*, and these variations induce in the secondary *s* the talking currents which pass over the line to the receiver R' of the operator. Although with this arrangement the primary and secondary currents pass simultaneously over the same line, speech is found to be transmitted quite as distinctly as when the two circuits are entirely separate, as is the case with the arrangement of Fig. 176. When the operator finds what number the subscriber wishes, she inserts the calling plug P' into the proper line and presses the ringing key k'. This cuts out the first subscriber, while the ringing is going on, by opening the contact points o' and closing the points o''. When the person called answers, the ringing key k' is released, and the two subscribers are thus connected and the magneto M, which actually runs all the time, is disconnected from both lines. Also, the operator releases her key k and thus cuts out her receiver while the conversation is going on. When one of the subscribers "hangs up", another lamp like l is lighted by a mechanism not shown here, and the operator then pulls out both plugs P and P'.

The bell b rings when an *alternating* p.d. is thrown upon the line, because, although the circuit is broken at t, an *alternating current* will surge into the condenser C and out of it and thus pull the armature first toward m and then toward n. The bell could not, of course, be rung by a direct current.

167. Wireless Telephone.* The principles employed in wireless telegraphy (radiotelegraphy) and telephone practice are combined in the wireless telephone. In Section 162 we saw that the detector had the property of transmitting current through it in one direction and that when the telephone was connected across the detector a musical note was heard whose frequency was the same (about 300 to 1,000) as that of the sending station, so that the receiver produced long or short notes for dashes and dots and these notes were of constant pitch. In wireless telephony the sending apparatus is so modified, that instead of sending out a succession of waves at equal intervals, it radiates a series of waves which correspond to the fundamentals and partials of the human voice. When this train of waves strikes the receiving station of any ordinary wireless station like that of Fig. 172 (b), it reproduces the human speech, since the diaphragm of the receiver is actuated by a series of currents the wave form of which corresponds to the wave form sent out by the sending station. Any wireless station, properly tuned, can "pick up" wireless telephone messages. Compared to wireless telegraphy, its operation is practical only over comparatively short distances and its uses as yet are quite limited.

* For one of several practical treatments of this subject, the reader is referred to "Experimental Wireless Stations", by Philip E. Edelmann, Minneapolis, Minnesota.

168. Automatic Enunciator. The transmitter, Fig. 179, of the automatic enunciator or loud speaking telephone is very similar to that used in the ordinary telephone. It differs in that

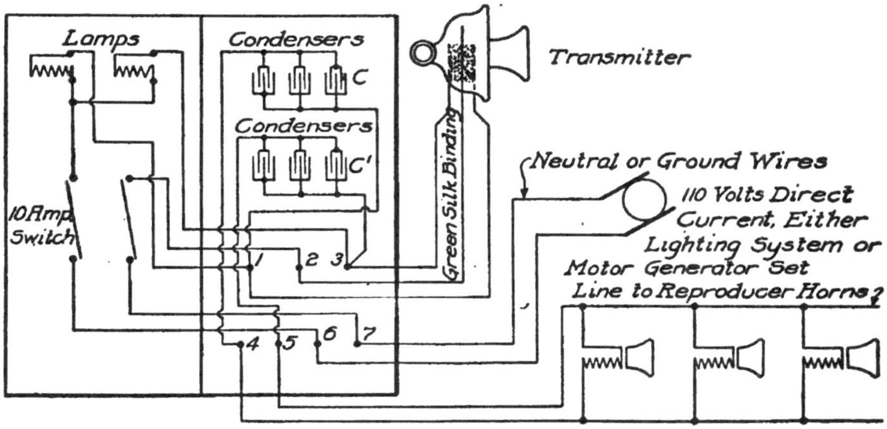

Fig. 179. Complete Automatic Enunciator System Showing Connections
Courtesy of Winkler and Reichmann, Chicago

the varying current through the transmitter passes from carbon disks placed both in front and in back of another carbon disk, to this latter disk, the disks being separated by granular carbon. One side of the transmitter is connected to one side of three condensers C

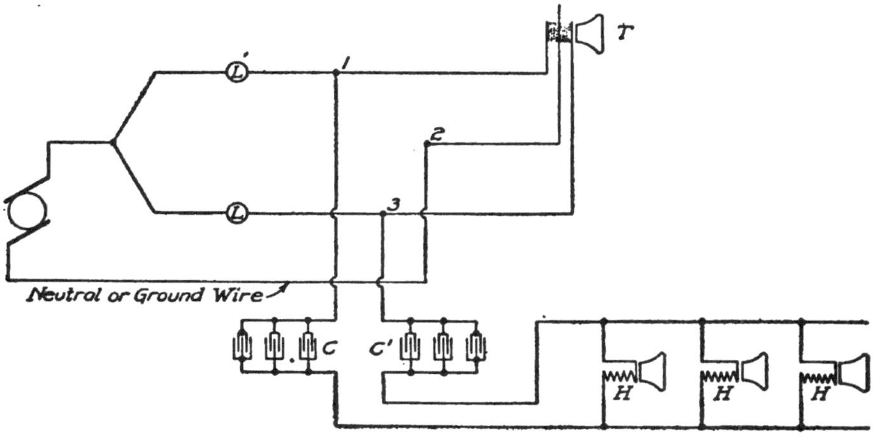

Fig. 180. Diagram of Automatic Enunciator System

in parallel, the other side of the three condensers running to one of the lines for the reproducer horns. Likewise, the other side of the transmitter is connected to one side of another three condensers C'

in parallel, the other side of these three condensers running to the other line for the reproducer horns.

When the sound vibrations from the voice strike the transmitter *T*, Fig. 180, a varying current through it results. This varies the potential of points *1* and *3* which are connected to the condenser system and consequently a varying potential is produced on the one side of the condenser systems *C* and *C′*. This varying potential induces a varying potential on the other side of the condensers to which the reproducer lines are attached, thus sending the talking current through the reproducers.

Fig. 181. Reproducer Complete with Horn
Courtesy of Winkler and Reichmann, Chicago

The complete reproducer is shown in Fig. 181 and its working parts are shown in Fig. 182. The reproducers are connected by the terminals *MM′* across the reproducer line, as many as fifty being connected in parallel in different parts of a building.

When the speaking current passes through the reproducer, it energizes the magnet and the iron core *C* acts upon the soft-iron armature *B* to which is attached the lever *E*. The lever *E* is attached to the lever *D*, and *D* to *F*, while *F* is attached to a light mica dia-

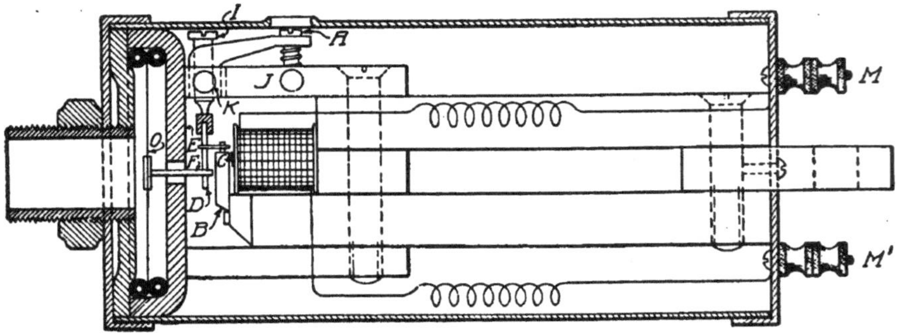

Fig. 182. Section Showing Working Parts of Reproducer
Courtesy of Winkler and Reichmann, Chicago

phragm *O*. Thus the motion of the armature due to the speaking currents is communicated by a system of levers to the mica diaphragm which reproduces the voice in any desired intensity depending upon the energy sent through the transmitter. For reproducing in a low voice, 10-watt lamps (high resistance) are placed at *L*, Fig. 180, in the transmitter circuit and for reproducing in a loud voice 60-watt lamps (low resistance) are used. *A*, *I*, and *K*, Fig. 182, are adjusting screws for the purpose of properly adjusting the air gap between *B* and *C*, and the lever system.

The automatic enunciator is used largely for quickly locating doctors, internes, and nurses in hospitals; for paging guests in hotels, members in clubs, etc.; for announcing trains in passenger terminals; for announcing umpires' decisions, and results in baseball parks and athletic fields, etc.

EXAMPLES FOR PRACTICE

1. Describe each of the following induction-coil discharge effects: (a) physiological, (b) heating, (c) mechanical, (d) chemical, (e) Gassiot's cascade, (f) in Geissler tubes, in X-ray tubes, (g) those applied to tuned circuits, Fig. 167, those applied to a wireless sending station, Fig. 171, and Fig. 172 (a).
2. Is the discharge through a Geissler tube intermittent or steady? How can this be shown? Give two ways.
3. What are cathode rays? Give two experiments which prove that they are negatively charged particles.
4. Give several differences between cathode rays and X-rays.
5. Describe how an X-ray picture of the hand is taken.
6. What is meant by a *radioactive* substance?
7. Describe Crookes' spinthariscope.
8. Describe the action of the coherer in a simple wireless outfit; of the detector in a modern wireless outfit.
9. What is the relation between the inductance and the capacity of the transmitting and receiving stations in a modern wireless outfit?
10. Explain the action of the simple telephone, Fig. 173.
11. Explain the use of the modern transmitter in connection with the telephone circuit (local-battery system, Fig. 176).
12. What is the difference in the train of waves sent out by a wireless-telegraph and a wireless-telephone sending station?
13. Can the same apparatus be used for receiving wireless-telephone messages that is used for receiving wireless-telegraph messages?

DYNAMO-ELECTRIC MACHINES

169. Classification. All dynamo-electric machines are commercial applications of Faraday's discovery of induced currents in 1831. They are all designed to transform the mechanical energy of a steam engine, a waterfall, a gasoline engine, etc., into the energy of an electric current. Whenever large currents are required—for example, in running street cars; in systems of lighting and heating; in the smelting, welding, and refining of metals; the charging of storage batteries, etc.—they are always produced by dynamo-electric machines.

There are two kinds of generators (1) d.c., or those producing a unidirectional (direct) current, that is, one which always flows in the same direction in the external circuit, and (2) a.c., or those producing an alternating current, that is, one which reverses in direction continuously throughout the entire circuit.

170. Principle of the A.C. Generator. The simplest form of a.c. generator consists of a single loop of wire in which an e.m.f. is generated by rotating the coil in a magnetic field, thus causing the wires to cut lines of force.

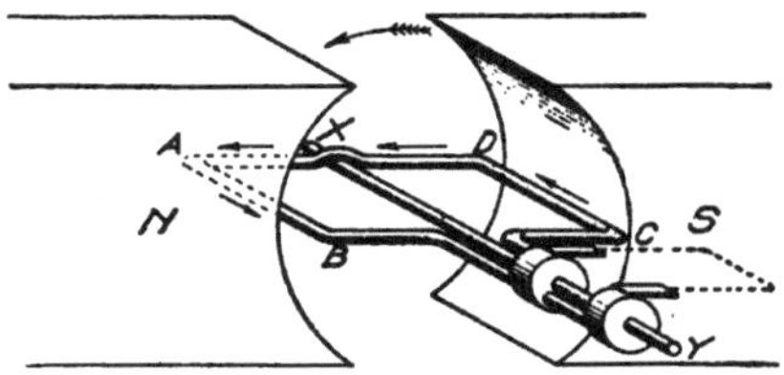

Fig. 183. Diagram of Simple A.C. Generator

The magnitude of the induced e.m.f. at any instant will, of course, depend upon the rate of cutting of lines of force, Section 130. Starting with the loop in the position shown in Fig. 183, and rotating the loop counter-clockwise about the axis xy will cause an induced current to flow in the direction indicated by the arrows according to Fleming's dynamo rule, Fig. 143, and rule at the end of Section 131. The e.m.f.'s induced in AB and CD for the position shown will have their maximum values since the wires are then cutting the magnetic flux at right angles and are consequently cutting more lines of force per second than in any other part of the revolution. Note that as CD moves up, AB moves down (and vice versa) across the magnetic flux so that the induced e.m.f.'s in all parts of the loop at any instant cause a current in the loop at that instant in one direction. As the coil moves from the position shown to a position 90 degrees farther on (one-fourth turn), the wires cut the magnetic flux at angles between 90 degrees and 0 degree so that the induced e.m.f. falls from a maximum to zero during this quarter turn. During the next quarter turn (90 degrees to 180 degrees) the wires are cutting the lines of force at angles varying from 0 degree to 90 degrees and the induced

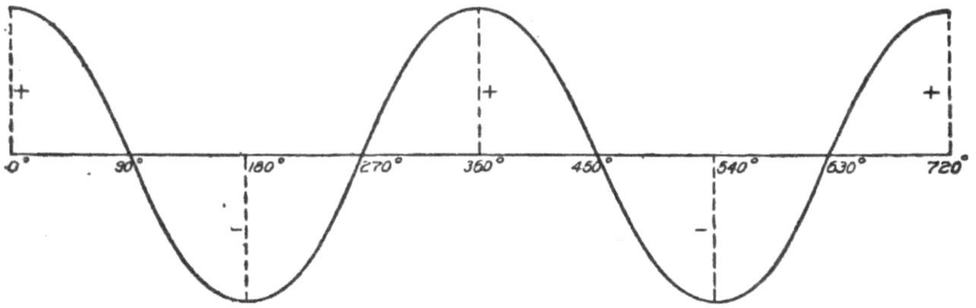

Fig. 184. E M.F. Curve of Simple A C Generator

e.m.f.'s change from zero to a maximum in the opposite direction. During the next quarter turn (180 degrees to 270 degrees) the wires are cutting the magnetic flux at angles varying from 90 degrees to 0 degree so that the induced e.m.f. falls from a maximum to zero but has been in the same direction as the quarter turn from 90 degrees to 180 degrees. During the next quarter turn (270 degrees to 360 degrees) the wires are cutting lines of force at angles varying from 0 degree to 90 degrees so that the induced e.m.f. varies from zero to a

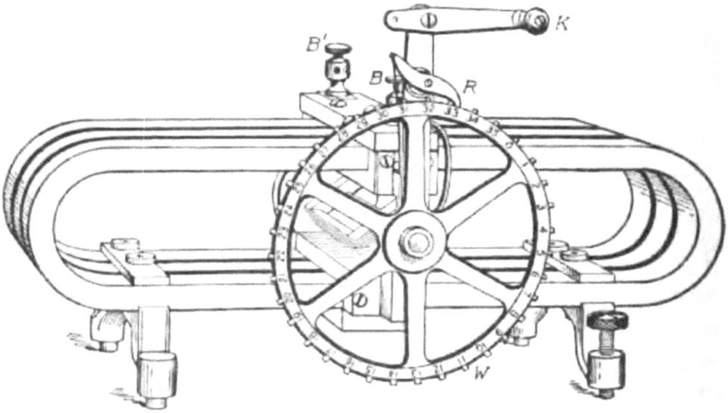

Fig 185. Simple Generator for Plotting E M F. Curve

maximum in the same direction as at the starting-point. It will be noted that the current changes direction (alternates) every time the plane of the loop is in the vertical position, since, as it passes through this position, the wires change from cutting up across the magnetic flux to cutting down across the magnetic flux, or vice versa. At this instant of change, the induced e.m.f. and current has zero value. Thus the current alternates twice during a revolution, reaching two maximum values which are in opposite directions, and two zero values.

These relations are shown by the curve in Fig. 184 for two revolutions (720 degrees) for a simple a. c. generator. If we consider the e.m.f. positive for the interval 0 degree to 90 degrees, then it would be negative for the interval 90 degrees to 270 degrees, positive again for the interval 270 degrees to 450 degrees, etc. Thus

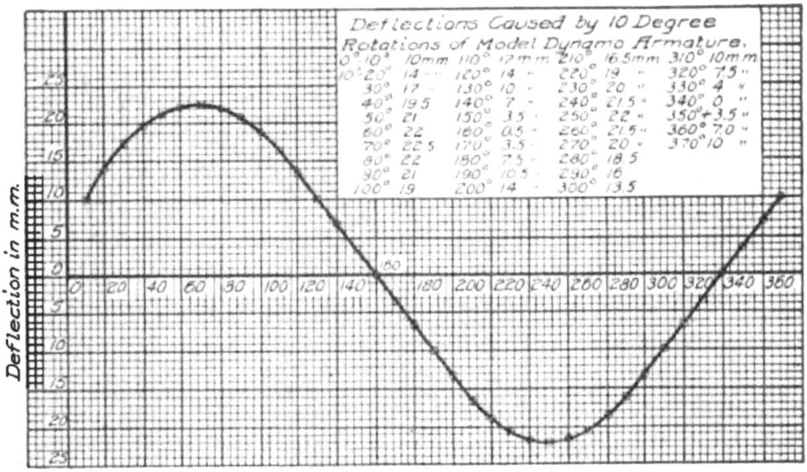

Fig. 186. E.M.F Curve Plotted from Observation with Apparatus Shown in Fig. 185

the e.m.f. alternates from one sign to the other and the current which it produces is called an *alternating current.*

An e.m.f. curve like that shown in Fig. 184 may be very beautifully obtained experimentally with the aid of the simple a. c. generator devised by Professor R. A. Millikan for this purpose and shown in Fig. 185. A coil of wire is rotated in a very uniform field produced by placing horseshoe magnets end-on, as shown in the figure. By rotating the crank *K* clockwise, the ratchet *R* is released and a spring rotates the coil through one of the ten-degree intervals into which the wheel *W* is divided by the stops placed around its circumference. As the coil rotates through one of these ten-degree intervals, an e.m.f. is induced which corresponds to the average angular displacement of the coil during this rotation of ten degrees. If a suitable galvanometer be put in circuit and the deflections thus obtained be plotted as ordinates, and the corresponding angular displacements of the coil as abscissas, the smooth curve shown in

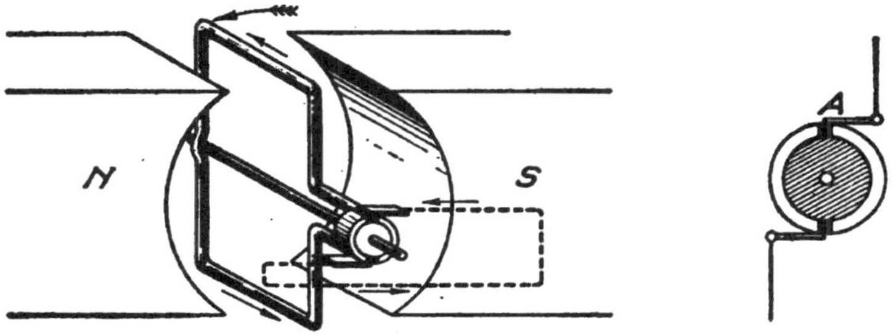

Fig. 187. Commutator Changes Alternating Current in Armature to Direct Current in External Circuit

Fig. 186 will result. Since the induced e.m.f.'s are proportional to the deflections, the curve has the exact form of the e.m.f. curve. This curve is theoretically a *sine wave* and made with this special apparatus it is *practically* a *sine wave.*

Commutator. By the use of the *commutator,* it is possible to transform a current which is alternating in the coils of the armature, to one which is a direct current, i.e., always flowing in the same direction in the external portion of the circuit. The simplest possible form of such a commutator is shown in Fig. 187 in connection with a single loop of wire, the simplest a.c. generator. The commutator consists of a metallic tube split into two semicircular segments and insulated from each other. One end of the rotating loop (or coil, if more turns are used as in Figs. 191 and 192) is soldered to one of these semicircles, and the other end to the other semicircle. The brushes

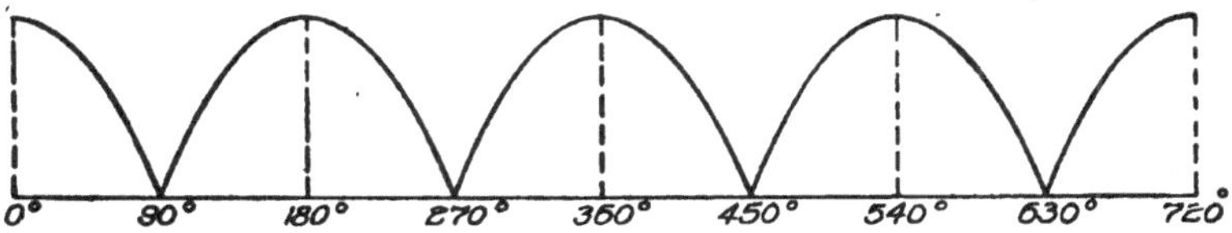

Fig. 188. E.M F. Curve with Commutator on Armature

are set in such a position that they lose contact with one semicircle and make contact with the other at the instant at which the current changes direction in the armature. The current therefore always passes out to the external circuit through the same brush, and is said to be a rectified, commutated, or unidirectional current. When a single loop or coil is used with the commutator, the current is not steady but pulsates with the e.m.f., rising to two maximum values and falling to zero value twice during each revolution. Thus the e.m.f. curve, Fig. 184, for the simplest single-loop a. c. generator becomes, when a commutator is used, the pulsating, unidirectional e.m.f. curve of Fig. 188.

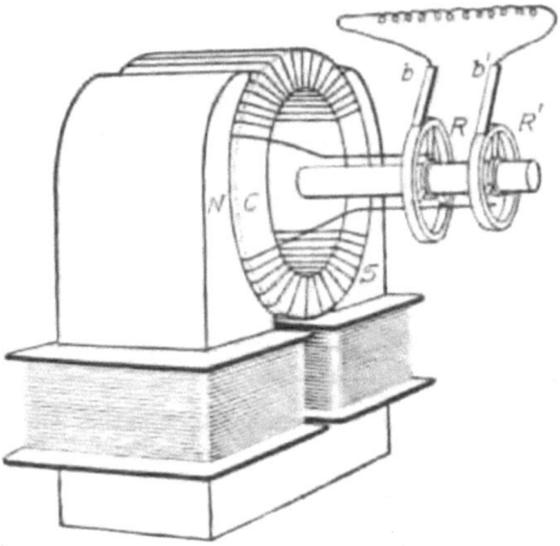

Fig. 189. Ring-Wound Armature

171. Simple A. C. Dynamo. The simplest form of commercial dynamo consists of a coil of wire so arranged as to rotate continuously between the poles of a powerful electromagnet, Fig. 189.

In order to make the magnetic field in which the conductor is moved as strong as possible, the coil is wound upon an iron core *C*. This greatly increases the total number of lines of magnetic force which pass between *N* and *S*, for the core offers an iron path, as shown in Fig. 190, instead of an air path from *N* to *S*.

The rotating part, consisting of the coil with its core, is called the *armature*. If the coil is wound in the manner shown in Figs. 189 and 190, the armature is said to be of the *ring* type.

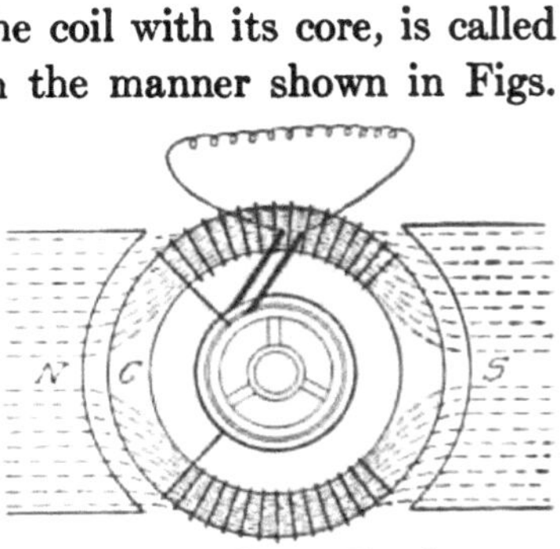

Fig 190. End View of Ring Armature

One end of the coil is attached to the insulated metal ring *R*, which is attached rigidly to the shaft of the armature and therefore rotates with it, while the other end of the coil is attached to a second ring *R'*. The brushes *b* and *b'*, which constitute the terminals of the external circuit, are always in contact with these rings, which rings are characteristic of all a. c. generators and are called the *collecting rings*.

As the coil rotates, an induced alternating current passes through the circuit. This current reverses direction as often as the coil passes through the position shown in Fig. 190, i.e., the position in which the conductors are moving *parallel* to the lines of force; for at this instant the conductors which have been moving up begin to move down, and those which have been moving down begin to

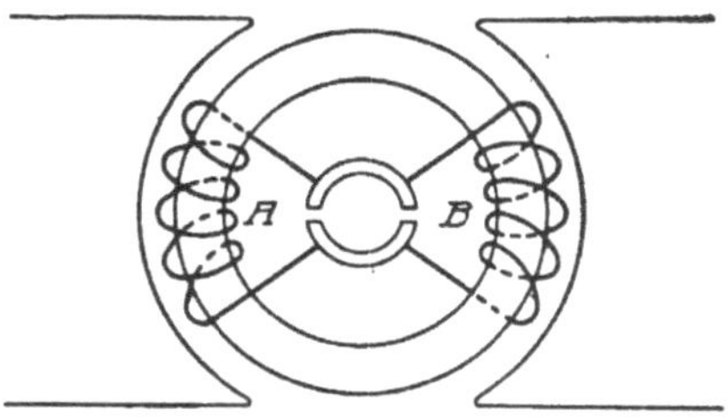

Fig. 191. Diagram of Two-Coil Armature

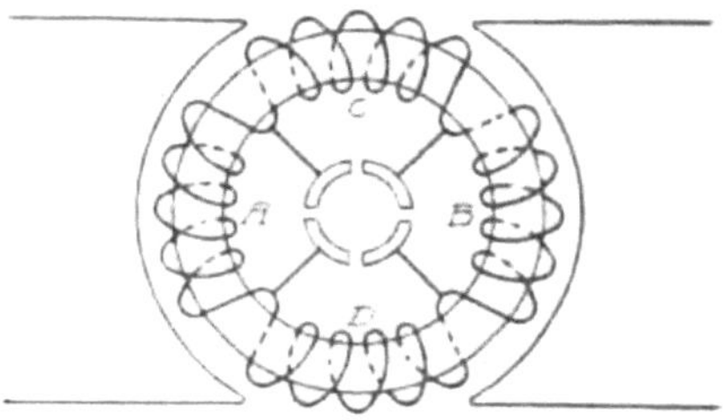

Fig 192. Diagram of Four-Coil Armature

move up. The current reaches its maximum value when the coils are moving through a position 90 degrees farther on than that shown in the figures, for then the lines of force are being cut most rapidly by the conductors on both sides of the coil. The e.m.f. curve of this generator is therefore like the one shown in Fig. 184.

172. Simple D.C. Dynamo. A simple two-coil armature used in connection with a commutator is shown in Fig. 191. The disadvantage here encountered is the pulsating character of the current, both coils reaching a maximum and minimum e.m.f., at the same time giving the e.m.f. curve of Fig. 188. This pulsating effect is avoided and a steady current produced in the commercial direct-current dynamo by building a commutator of a large number of segments and connecting the segments across small portions of the armature coil. Even a four-coil armature with four segments on the

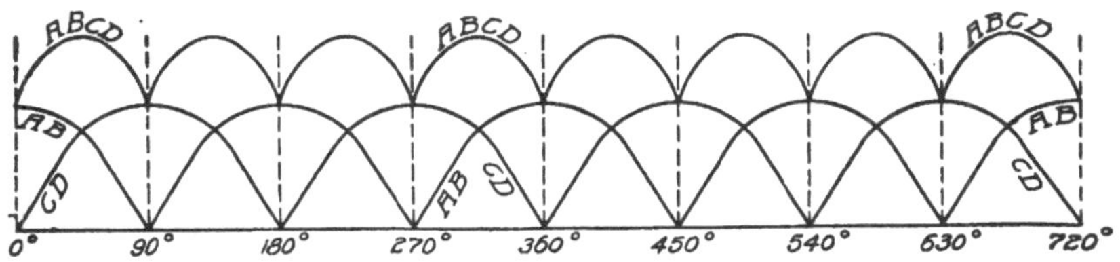

Fig. 193. E.M F. and Current Curve for Four-Coil Armature

commutator will, in a large measure, overcome the pulsating effect of the current as can be seen from the following considerations. Starting with the coils *A*, *B*, *C*, and *D*, Fig. 192, in the positions shown, the e.m.f. induced in the coils *A* and *B* will be a maximum at this instant shown by the e.m.f. curve marked AB in Fig. 193, and the e.m.f. induced in the coils C and D at the same instant will be zero. A quarter of a turn later, the e.m.f. in A and B will be zero and in C and D will be a maximum. Since the e.m.f. curve for the external portion of the circuit is the sum of these or the curve $ABCD$, it will be seen that at no place does this curve fall to zero; this means that the pulsations of current are smaller than they would be for a single loop, the current of this four-coil armature being very nearly steady.

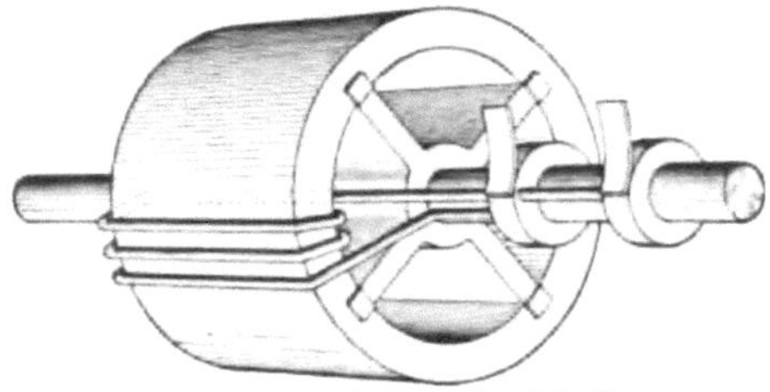

Fig. 194. Single Gramme-Ring Winding

The four coils of Fig. 192 are in series and may be thought of as one coil with four connecting wires running to the four segments of the commutator. In practice, the coil is thus divided into a very large number of parts, a hundred or more, so that some portion of the coil is always having a maximum e.m.f. induced in it, thus making the undulations (pulses) which are superimposed on the steady current so small that some special device such as the telephone must be used to detect them. Fig. 194 shows a single gramme-

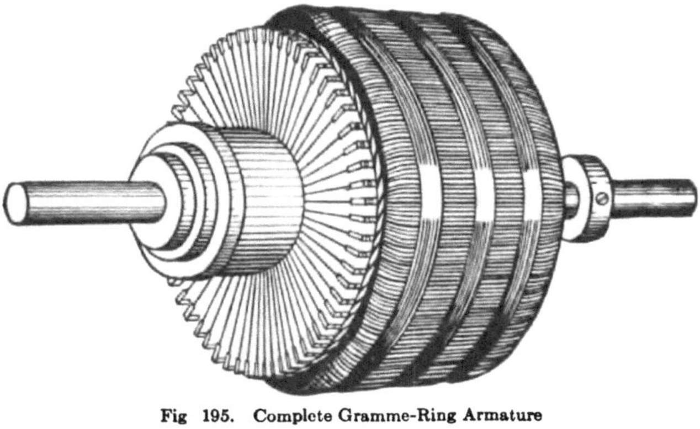

Fig 195. Complete Gramme-Ring Armature

ring winding. The ring as will be seen in the diagram is really a hollow cylinder, which in all the previous diagrams has been shown only in cross-section. Fig. 195 is a typical commercial gramme-ring armature. The commutator is not always built on next to the shaft but is sometimes built around the periphery of the armature itself so that the outside of the ring forms the commutator.

DIRECT-CURRENT GENERATORS

173. Ring-Armature Direct-Current Dynamo. Fig. 196 is a diagram illustrating the construction of a commercial two-pole direct-current dynamo of the ring-armature type. The figure represents an end view of a core like that shown in Fig. 189. The coil is wound continuously around the core, each segment being connected to a corresponding segment of the commutator, in the manner shown in the figure. At a given instant, currents are being induced in the same direction in all the conductors on the outside of the core on the left half of the armature. The cross on these conductors, representing the tail of a retreating arrow, is to indicate that these currents flow away from the reader. No e.m.f.'s are induced in the conductors on the inner side of the ring, since these conductors cut no lines of force, Fig. 190; nor are currents induced in the conductors at the top and bottom of the ring where the motion is parallel to the magnetic lines. The currents from the two halves of the ring pass out at b through the line and back at b'. This condition always exists, no matter how fast the rotation; for it will be seen that as each loop rotates into the position where the direction of its current reverses

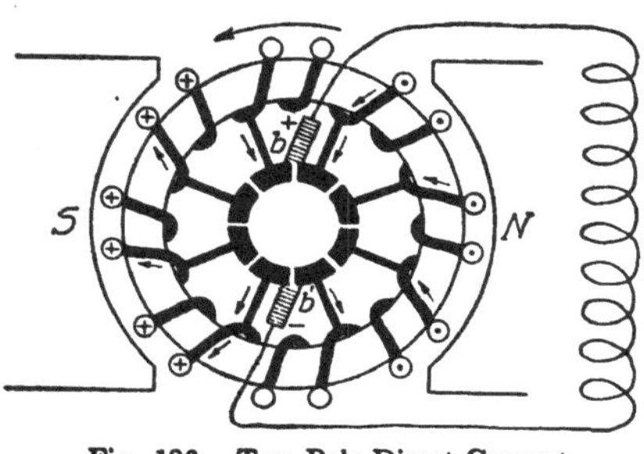

Fig. 196. Two-Pole Direct-Current Dynamo Diagram

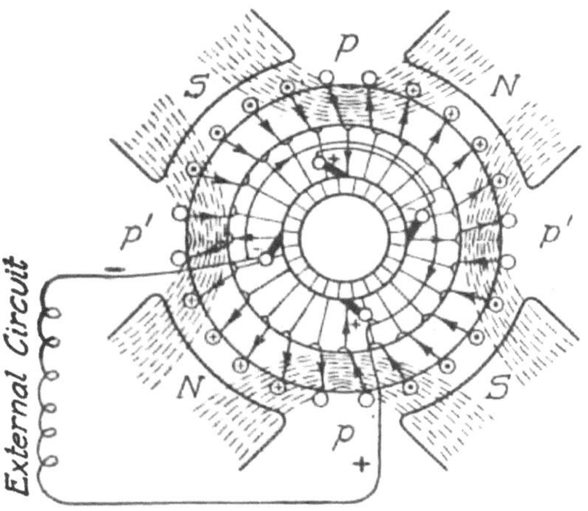

Fig. 197. Four-Pole Direct-Current Dynamo Diagram Showing Direction of Lines of Force

it passes a brush and therefore at once becomes a part of the circuit on the other half of the ring where the currents are all flowing in the opposite direction.

If the machine is of the four-pole type, like that shown in Fig. 197, the currents flow toward two neutral points, *pp*, or points of no induction, instead of toward one, as in two-pole machines. Hence there are four brushes, two positive and two negative, as in the figure. Since the two positive and the two negative brushes are connected as shown, both sets of currents flow off to the external circuit on a single wire. The figure with its arrows will explain completely the generation of currents by a four-pole machine.

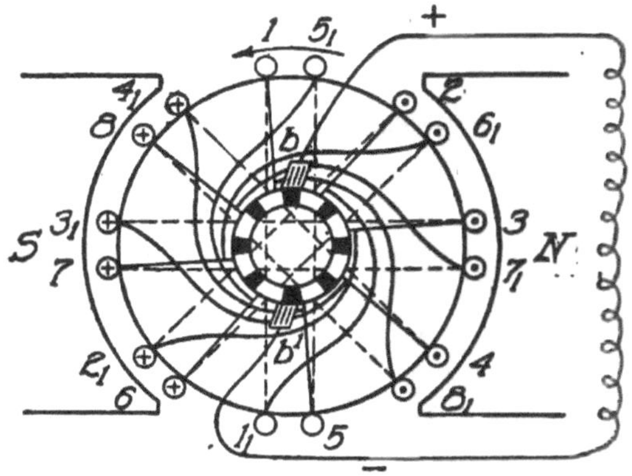

Fig. 198. Two-Pole Direct-Current Generator with Drum Winding

174. Drum-Armature Direct-Current Dynamo. The drum-wound armature, shown in section in Fig. 198, has an advantage over the ring armature in that, while the conductors on the inside of the latter never cut lines of force and are, therefore, always idle, in the former all of the conductors are cutting lines of force except when they are passing the neutral points. In theory, however, the operation of the drum armature is precisely the same as that of the ring armature. All the conductors on the left side of the line connecting the brushes, Fig. 198, carry induced currents which flow in one direction, while all the conductors on the right side of this line have opposite currents induced in them. It will be seen, however, in tracing out the connections *1*, 1_1, *2*, 2_1,

Fig. 199. Modern Four-Pole Direct-Current Generator with Interpoles

3, 3_1, etc., of the figure (the dotted lines representing connections at the back of the drum), that the coil is so wound about the drum that the currents in both halves are always flowing toward one brush *b*, from which they are led to the external circuit. Fig. 199 shows a typical modern four-pole generator, and Fig. 200 the corresponding drum-wound armature. Fig. 156 illustrates nicely the method of winding such an armature, each coil beginning on one segment of the commutator and ending on the adjacent segment.

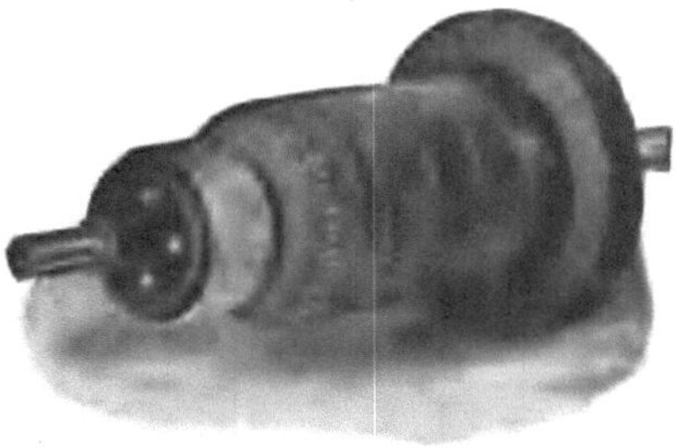

Fig. 200. Drum-Wound Armature for Generator Shown in Fig. 199

175. Methods of Exciting Generator Fields and Their Applications. *Magneto.* To generate an e.m.f., conductors must cut across a magnetic flux. In the magneto, the magnetic flux is produced by permanent magnets. Since permanent magnets are much bulkier than electromagnetic ones of the same strength, it is impractical to use permanent magnets on large genera-

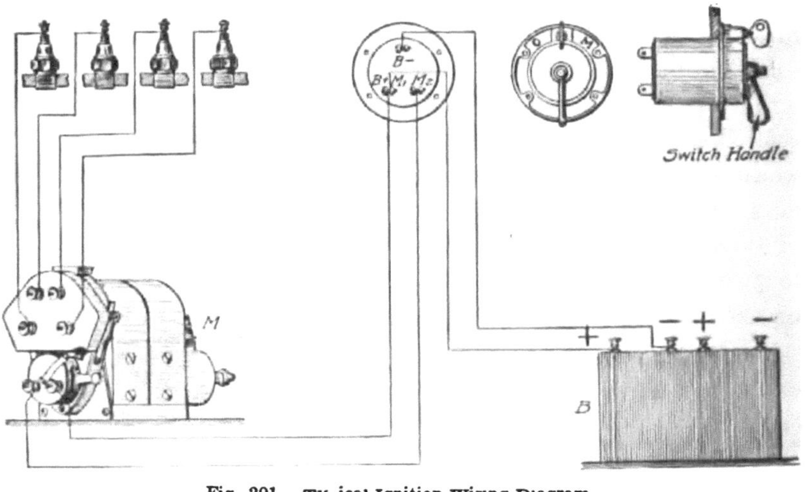

Fig. 201. Typical Ignition Wiring Diagram
Courtesy of Thomas B. Jeffery Company, Kenosha, Wisconsin

tors on account of the massive magnets which would have to be used. For this reason, they are used only where a small amount of power is required, for example, in the ringing of bells on some telephone systems and in producing the sparks for ignition in internal combustion engines. Fig. 201 shows how the magneto M is connected for ignition purposes to the four cylinders of an automobile. The car is usually started by using the spark from a set of about four dry cells B and then the switch is thrown over to connect the spark plugs to the magneto.

Separately Excited Field. Here the magnetic flux is produced by an electromagnet, the current for which is furnished either by a set of cells or another d.c. generator, Fig. 202. This system of field excitation is seldom used in direct current work except when voltages greater than about 800 volts are to be obtained, or when a great many generators are in operation at once. It is, however, very largely used for alternating current generators, since a unidirectional magnetic flux must be produced, and the alternating current will not produce it.

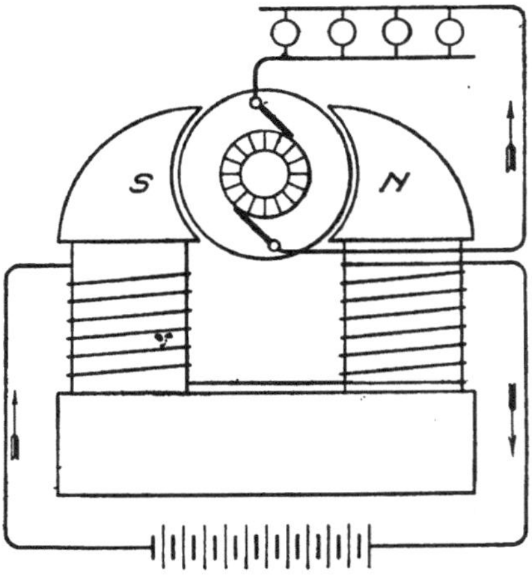

Fig 202. Direct-Current Generator with Separately Excited Field

The magnitude of the induced e.m.f. in this type of generator is controlled by varying the current which flows around the field magnets, or else by varying the number of turns of the solenoid of the field magnets. These would naturally be the means to employ, since the strength of a magnet depends on the ampere-turns in its circuit. Its e.m.f. might also be controlled by changing the speed of rotation of the armature, but this is not found to be a practical way. It can also be controlled by changing the point of commutation, that is, by rotating the brushes forward in the direction of rotation to a point beyond that at which commutation of the current in the

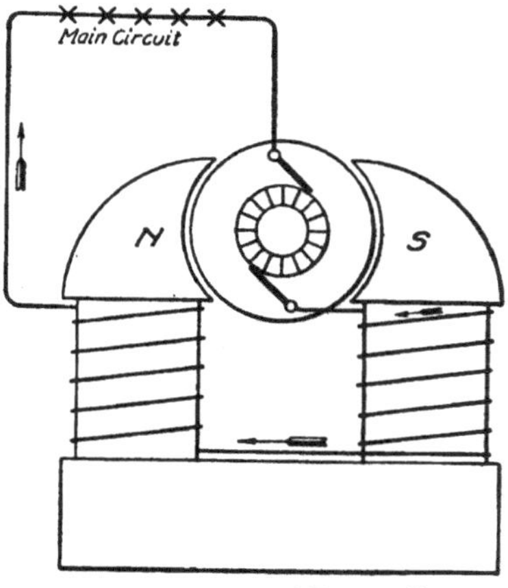

Fig 203. Series-Wound Direct-Current Generator

coil takes place. The latter method is impractical since it produces bad sparking at the brush contact with the commutator.

Self-Excited Fields. Series Winding. In almost all direct-current machines, the field magnet *NS*, Fig. 203, is excited by the current which the dynamo itself produces. In all self-exciting machines there is enough residual magnetism left in the iron cores after stopping, to start feeble induced currents when started up again. These currents immediately increase the strength of the magnetic field, and so the machine quickly builds up its current until the limit of magnetization is reached. In the series-wound dynamo, Fig. 203, the windings on the field magnet consist of a few turns of heavy wire in series with the "main circuit" so that *all* of the current generated flows around the solenoids of the field magnet. A decrease in the external resistance would cause more current to flow, producing a stronger field, so that if the dynamo were driven at constant speed, the e.m.f. would be increased. Likewise, an increase in the external resistance would cause a decrease in the e.m.f. produced by the dynamo. In practice, arc lighting is the only use for which series-wound dynamos are employed, since for any system of arc lights the voltage and load are both constant. About fifty lamps are commonly fed by one machine. This requires a dynamo capable of producing a voltage of 2500 volts, since each lamp requires a pressure of about 50 volts. Since an arc lamp requires a current of about 10 amperes, such a dynamo must have a power of $2500 \times 10 = 25{,}000$ watts, or 25 kilowatts. This is equivalent to about 33.5 horsepower.

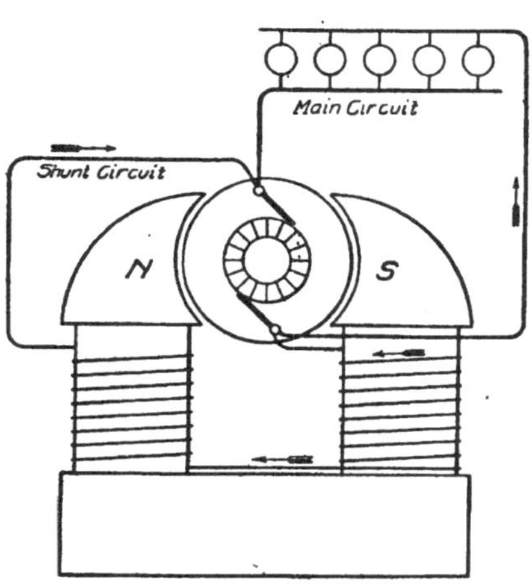

Fig. 204. Shunt-Wound Direct-Current Generator

Shunt Winding. In the so-called shunt-wound machines a small portion of the current is led off from the brushes through the shunt circuit, Fig. 204, which consists of many turns of fine wire which encircle the core of the field magnet, while the rest of the current flows through the main circuit. Since the shunt circuit and the main circuit are in parallel, the resistance of the shunt circuit must be high compared with that of the main circuit. The ampere-turns encircling the field magnet are the same as in a similar-series dynamo in order to produce a given magnetic flux. The magnetic flux may be increased by decreasing the resistance in the shunt circuit, thus increasing the magnetizing current. This dynamo, if properly designed, has a fairly constant e.m.f. within certain limits, but if overloaded, that is, if too large currents are drawn from it, the external resistance under these conditions will be too small so that not enough current will flow through the shunt circuit to energize the magnets and the e.m.f. generated will drop to zero. Thus we have seen that increasing the load on a series dynamo increases the e.m.f. and on a shunt-wound dynamo decreases the e.m.f. A combination of these two windings, then, in the proper way should enable us to get a constant e.m.f. for any load. This is accomplished by the aid of the so-called *compound winding*.

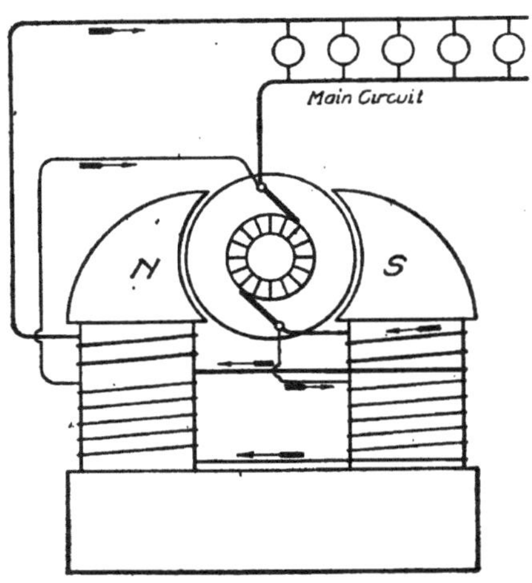

Fig. 205. Compound-Wound Direct-Current Generator

Compound Winding. In the compound-wound dynamo, Fig. 205, the field magnet is energized by the current from a shunt circuit like that in the shunt-wound dynamo. It is also energized by the current which flows in the main circuit which is led around the field magnet. This latter set of windings is called the *series* winding since they are in series with the main circuit. When the load on this dynamo is increased, that is, when the resistance of the main circuit is made less, this larger current flows around the field magnet thus producing a stronger magnetic flux for the armature to cut, and thus preventing the e.m.f. from decreasing as in the case of the shunt-wound motor. By properly proportioning the ampere-turns in the shunt winding and the series winding, it is possible to have the e.m.f. produced by this compounding remain constant, decrease, or increase, as the dynamo load is changed from no load to full load. Hence it is that the compound-wound dynamo is always used for incandescent-lighting systems and wherever a variable load at constant voltage is to be carried by the dynamo.

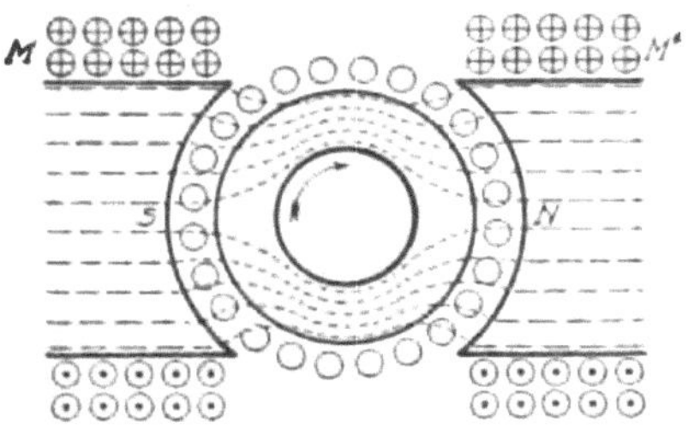

Fig. 206. Flux Distribution with Field Only Excited

176. Importance of Setting Brushes to Agree with Point of Commutation. In our preceding discussion of the commutator, Section 170, we assumed that commutation (reversing of current) took place when the plane of the coil was in the position *PP*, Fig. 208. In practice, this is found not to be true and the reason can readily be seen from the following considerations: In Figs. 206, 207, and 208, the dots on a conductor (heads of approaching arrows) indicate currents flowing toward the observer while the crosses (tails of receding arrows) indicate currents flowing away from the observer. The flux distribution through a ring armature, when the current in the field coils *MM'* is on, is shown in Fig. 206 (no current is flowing in the armature). The flux distribution when the current is flowing in the armature and none in the field coils is shown in Fig. 207. When the current is flowing in both the armature and the field coils, that is, when the dynamo is carrying a load, the flux distribution is like that shown in Fig. 208. It will be noticed that the flux distribution in the dynamo is the same as it would be if the magnetic flux were rotated somewhat in the direction of rotation of the armature. Hence the position of the coil when the current in it reverses is not in the plane *PP* but a plane *P'P'*, a little in

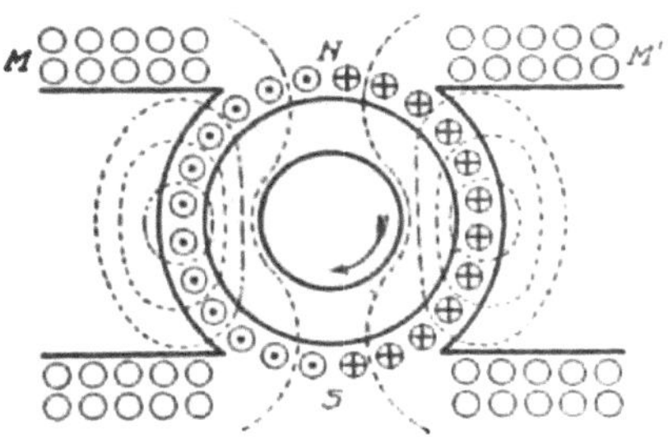

Fig. 207. Flux Distribution with Armature Only Excited

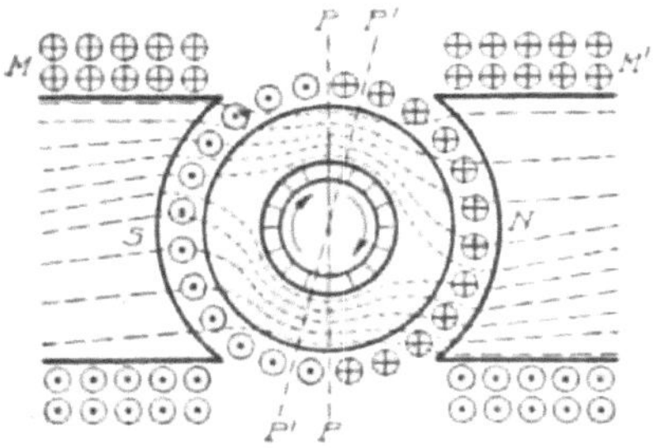

Fig. 208. Flux Distribution in Generator Showing Proper Position of Brushes

advance of PP. The brushes, therefore, should be set to make contact with the commutator in the plane $P'P'$. In the case of the motor, however, the reverse is true, so that the brushes, instead of being rotated forward, should be rotated backward. A failure to thus set the brushes on dynamos and motors is often a serious source of sparking* at the brushes. In the case of drum-wound armatures the amount which the brushes need to be rotated is much less, because the self-induction of the armature is less and hence commutation takes place more quickly.

EXAMPLES FOR PRACTICE

1. Diagram and explain a simple a.c. generator. Draw the e.m.f. curve for such a generator.

2. Explain with the aid of a diagram how the commutator changes the current in the external circuit from an alternating to a direct current.

3. With the aid of the dynamo rule, explain why, in Fig. 197, the current in the conductors under the south poles is moving toward the observer, and that in the conductors under the north poles away from the observer. Explain in a similar way the direction of the currents in Figs. 196 and 198.

4. If a direct-current dynamo of the type shown in Fig. 197 had ten poles, how many brushes would the commutator require?

5. A ring armature which develops the same e.m.f. as a drum armature has nearly twice as much wire and therefore nearly twice as much resistance. Why?

6. If a series-wound dynamo is running at a constant speed, what effect will be produced on the strength of the field magnets by diminishing the external resistance and thus increasing the current? What will be the effect on the e.m.f.? (Remember that the whole current goes around the field magnets.)

7. If a shunt dynamo is run at constant speed, what effect will be produced on the strength of the field magnets by reducing the external resistance? What effect will this have on the e.m.f.? (Remember that reducing the external resistance causes a smaller fraction of the current to flow through the shunt.)

8. In an incandescent-lighting system the lamps are connected

*For other causes and methods of preventing sparking during commutation the reader is referred to the literature on the subject of "Dynamo Electric Machinery".

in parallel across the mains. Every lamp which is turned on, then, diminishes the external resistance. Explain from a consideration of Examples 6 and 7 why a compound-wound dynamo keeps the p.d. between the mains constant.

9. Single dynamos often operate as many as 10,000 incandescent lamps at 110 volts. If these lamps are all arranged in parallel and each requires a current of .5 ampere, what is the total current furnished by the dynamo? What is the activity of the machine in kilowatts and in horsepower?

10. How many 110-volt lamps can be lighted by a 12,000-kilowatt generator?

11. Should the brushes on a dynamo be rotated forward or backward to make contact at the instant of commutation in the coils? Is the same true with regard to setting the brushes on a motor?

12. Why does it take twice as much work to keep a dynamo running when 1,000 lights are on the circuit as when only 500 are turned on?

DIRECT-CURRENT MOTORS

177. Dynamo Used as a Motor. In construction the motor* differs in no essential respect from the dynamo. To analyze the operation, as a motor, of such a machine as that shown in Fig. 196, suppose a current from an outside source is first sent around the coils of the field magnets and then into the armature at b'. Here it will divide and flow through all the conductors on the left half of the ring in one direction, and through all those on the right half in the opposite direction. Hence, in accordance with the motor rule, all the conductors on the left side are urged upward by the influence of the field, and all those on the right side are urged downward. The armature will therefore begin to rotate, and this rotation will continue so long as the current is sent in at b' and out at b. For as fast as coils pass either b or b', the direction of the force acting on them changes. The left half is therefore always urged up and the right half down. The greater the strength of the current, the greater the force acting to produce rotation.

If the armature is of the drum type, Fig. 198, the conditions are not essentially different. For, as may be seen by following out the

*Before reading the following pages on d. c. motors, the student should familiarise himself with the contents of Articles 133 and 134, especially the motor rule in Section 133.

windings, the current entering at b' will flow through all the conductors on the left half in one direction and through those on the right

Fig. 209. Allis-Chalmers Interpole Railway Motor

half in the opposite direction. The commutator keeps these conditions always fulfilled. The analysis of the operation of a four-pole dynamo, Fig. 197, as a motor is equally simple.

178. Street-Car Motors. Electric street cars are nearly all operated by direct-current series-wound motors placed under the cars and attached by gears to the axles. Fig. 209 shows a typical street-car motor of the compensating-pole type. The upper field poles with their interpoles are raised with the case when the motor is opened for inspection, as in the figure. The current is generally supplied by compound-wound dynamos which maintain a constant potential of about 500 volts between the trolley, or third rail, and the track which is used as the return circuit. The cars are always operated in parallel, as shown in Fig. 210. In a few instances, street cars are operated upon alternating- instead of upon direct-current circuits. In such cases the motors are essentially the same as direct-current series-wound motors; for since in such a machine the current must reverse in the field magnets at the same time that it reverses in the armature, it will be seen that the armature is always impelled

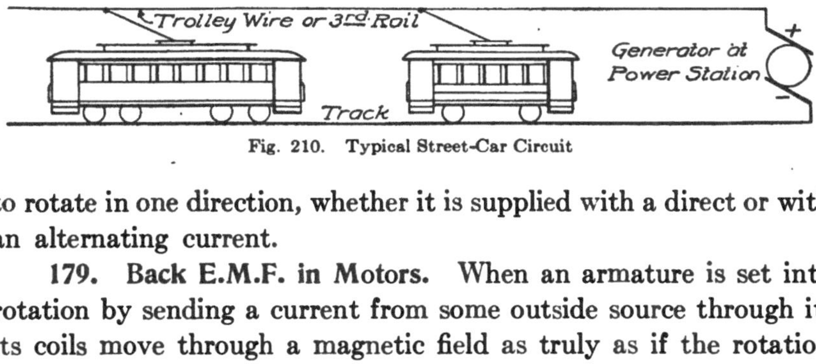

Fig. 210. Typical Street-Car Circuit

to rotate in one direction, whether it is supplied with a direct or with an alternating current.

179. Back E.M.F. in Motors. When an armature is set into rotation by sending a current from some outside source through it, its coils move through a magnetic field as truly as if the rotation were produced by a steam engine, as is the case in running a dynamo. An induced e.m.f. is therefore set up by this rotation. In other words, while the machine is acting as a motor, it is also acting as a dynamo. The direction of the induced e.m.f. due to this dynamo effect will be seen from Lenz's law, or from a consideration of the dynamo and motor rules, to be opposite to the outside p.d. which is causing current to pass through the motor. The faster the motor rotates, the faster the lines of force are cut, and hence the greater the value of this so-called *back* e.m.f. If the motor were doing no work, the speed of rotation would increase until the back e.m.f. reduced the current to a value sufficient simply to overcome friction. It will be seen, therefore, that, in general, the faster the motor goes, the less the current which passes through its armature, for this current is always due to the *difference* between the p. d. applied at the brushes—500 volts in the case of trolley cars—and the back e.m.f. When the motor is starting, the back e.m.f. is zero, and hence, if the full 500 volts were applied to the brushes, the current sent through would be so large as to ruin the armature through overheating. To prevent this, each car is furnished with a controller (starting-box,

Fig 211. Typical Street-Car Control Apparatus

Fig. 211) which enables the motorman to throw resistance into series with the two motors with which a car is usually equipped. As the controller lever is rotated, this resistance is cut out of the circuit leaving the two motors in series, each receiving one-half of the line voltage and therefore running at half-speed. The next position of the controller lever either shunts the fields or short-circuits part of them, again increasing speed. Further rotation removes the shunt and connects the motors in parallel, but with a resistance in the circuit which is again cut out by further rotation of the controller lever, leaving each motor to receive the full-line voltage. Finally the last position of the controller either shunts the fields or short-circuits part of them, and the motor is then running at full speed.

The series-wound motor is also used on electric automobiles and other motor vehicles. These motors have a large starting force, proportional to the *square* of the current, whereas in shunt-wound motors with constant fields, the starting force is proportional to the current sent through the armature.

180. Shop Motors. The shunt-wound motor is the one most commonly used in shops on account of the fact that its speed is nearly

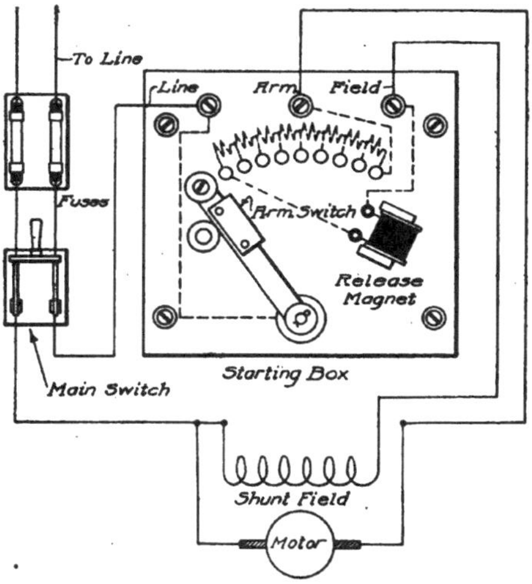

Fig. 212. Wiring Connections for Starting Box of a Shunt-Wound Motor

constant for all loads. Fig. 212 shows the wiring connections for the starting box of such a motor. If it is desirable to vary the speed of the motor, this may be accomplished by placing a variable resistance in series with the field and when this is done, the ordinary shunt-wound motor may satisfactorily be made to vary about thirty per cent in speed.

In starting the motor, the main switch, Fig. 212, is first closed, then the arm switch is slowly moved to the right, thus gradually cutting out the starting resistance as the motor acquires its running speed. The arm switch is held in position then by the release magnet until the motor is again stopped by opening the main switch, when the current in the field gradually falls to zero and the release magnet weakens, allowing the spring in the arm switch to throw it to the "off" position.

In the running of textile machinery, where a very constant speed for variable loads is required, the differential motor is used, which is nothing but a compound dynamo used as a motor. For operating

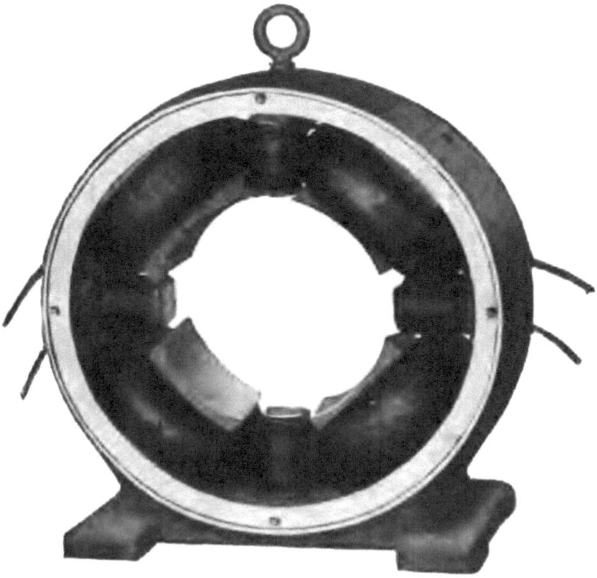

Fig. 213. Field Frame and Coils of a Typical Interpole Motor

derricks, cranes, and machinery requiring a large starting force (starting torque), the series-wound motor is used.

181. Reversing Motor. The reversing of a motor is accomplished, first, by stopping the motor, then, by putting resistance in the line, and finally, by reversing the current through the armature of the motor by means of a reversing switch. Since the position of the brushes for "no-sparking" commutation is not the same for a load as for the "no-load" position, Section 176, it is evident that to set the brushes properly for the motor to run in one direction would make them spark very badly when the motor was reversed; hence for a reversing motor the brushes should be set for the no-load position, i.e., in the plane *PP*, Fig. 208. The so-called commutating pole used on interpole dynamos, Fig. 213, and motors produces a magnetic field which aids the current in reversing and prevents sparking through a very wide range of load. It also allows the field to be made very much weaker than ordinary motors and thus the range of speed can be made 4 to 1 by using control resistance in series with the field. Since sparking during commutation is elim-

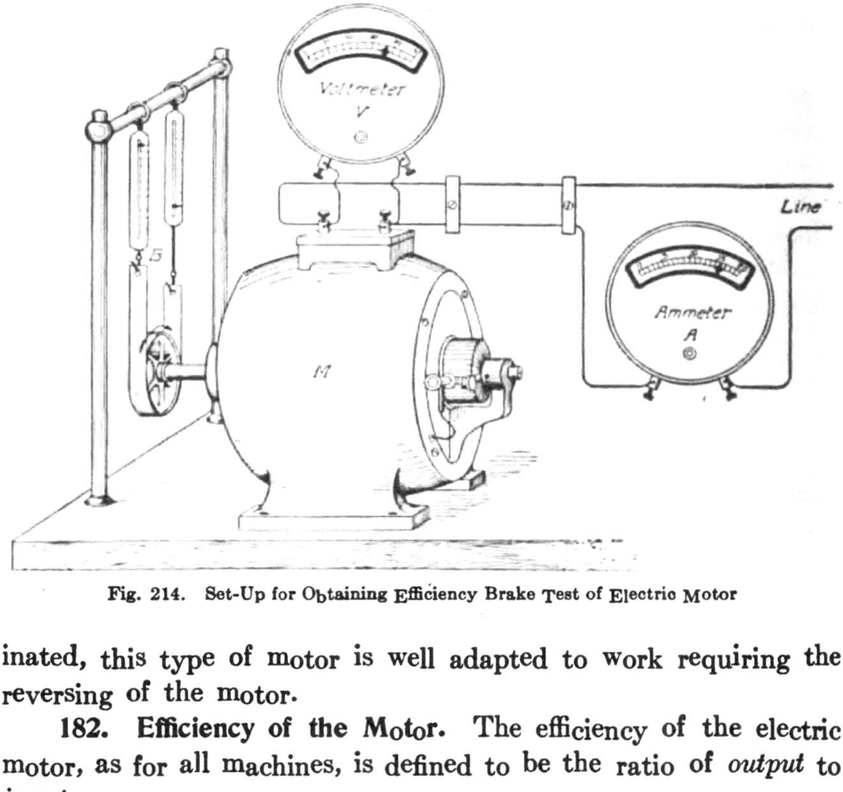

Fig. 214. Set-Up for Obtaining Efficiency Brake Test of Electric Motor

inated, this type of motor is well adapted to work requiring the reversing of the motor.

182. Efficiency of the Motor. The efficiency of the electric motor, as for all machines, is defined to be the ratio of *output* to *input*, or

$$\text{efficiency} = \frac{\text{output}}{\text{input}}$$

A simple form of brake, Fig. 214, consists of a belt passing under the pulley on the motor shaft and connected to the spring balances *B*. The friction between the belt and the pulley will cause one balance-reading to increase and the other to decrease, when the motor is running, so that the pull on the belt equals the difference between the two balance-readings, or in the case of the modified Prony brake, Fig. 215, it is equal to the single balance-reading minus the weight W. The pull on the belt in pounds multiplied by the circumference of the pulley in feet gives the number of foot-pounds of work done by the motor in one revolution. The work per revolution multiplied by the r.p.m. (revolutions per minute) gives the work done per minute which, divided by 33,000, gives the horsepower of the motor (1 h.p. = 33,000 foot-pounds per minute).

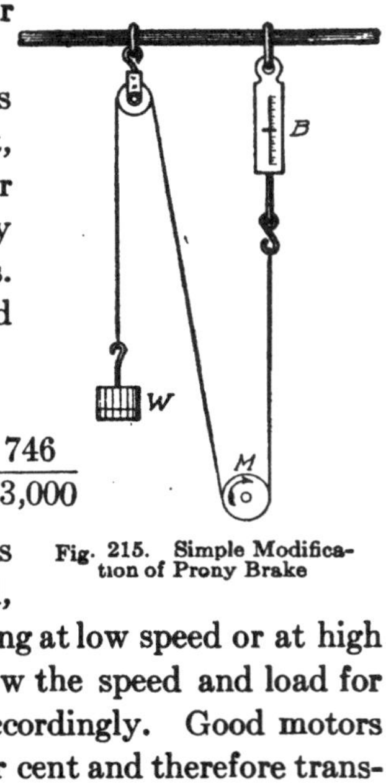

Fig. 215. Simple Modification of Prony Brake

The *input*, or rate at which energy is supplied to the motor by the electric current, expressed in watts, is equal to the number of volts p.d. across the motor multiplied by the current through the motor in amperes. The *input* in *horsepower* equals watts divided by 746 (1 h.p. = 746 watts). Therefore

$$\text{efficiency} = \frac{\text{ft. lb. per revolution} \times \text{r.p.m.}}{\text{volts} \times \text{amperes}} \times \frac{746}{33{,}000}$$

A motor can be designed to have its highest efficiency when carrying a small load, or when carrying a large load, or when running at low speed or at high speed. Hence, it is very important to know the speed and load for which it is most efficient and to design it accordingly. Good motors have efficiencies from 75 per cent to 90 per cent and therefore transform electrical energy into mechanical work with but little loss of energy. This fact coupled with their safety and convenience is rapidly transforming shops from a confusion of belts and dangerous transmission shafting to safe and orderly motor-equipped shops.

EXAMPLES FOR PRACTICE

1. A current is flowing from top to bottom in a vertical wire. In what direction will the wire tend to move on account of the earth's magnetic field?

2. If a current is sent into the armature of Fig. 196 at b', and taken out at b, which way will the armature revolve?

3. When an electric fan is first started, the current through

it is much greater than it is after the fan has attained its normal speed. Why?

4. If, in the machine of Fig. 197, a current is sent in on the wire marked +, what will be the direction of rotation?

5. Would an armature wound on a wooden core be as effective as one made of the same number of turns wound on an iron core?

6. Will it take more work to rotate a dynamo armature when the circuit is closed than when it is open? Why?

7. Show that if the reverse of Lenz's law were true, a motor once started would run of itself and do work, i.e., it would furnish a case of perpetual motion.

8. Explain why a series-wound motor can run either on a direct or an alternating circuit.

9. If the pressure applied at the terminals of a motor is 500 volts, and the back pressure, when running at full speed, is 450 volts, what is the current flowing through the armature, its resistance being 10 ohms?

10. An electric automobile is run for five hours. During this time the motor delivers energy at an average rate of 2 h.p. If the motor has an efficiency of 90 per cent and the storage batteries an efficiency of 75 per cent, how much does it cost to charge the storage batteries sufficiently for this trip, if the cost of the electricity used in charging the batteries is 4 cents per kilowatt hour?

ALTERNATING-CURRENT GENERATORS

183. Multipolar Alternator. For most commercial purposes it is found desirable to have 120 or more alternations of current per second. This could not be attained easily with two-pole machines like those sketched in Figs. 183, 189, and 191. Hence, commercial alternators are usually built with a large number of poles alternately *N* and *S*, arranged around the circumference of a circle in the manner shown in Fig. 216. The dotted lines represent the direction of the lines of force through the iron. It will be seen that the coils which are passing beneath *N* poles have induced currents set up in them, the direction of which is opposite to that of the currents which are induced in the conductors passing beneath *S* poles. Since, however, the direction of the winding of the armature coils changes between each two poles, all the inductive effects of all the poles are added together in the coil and constitute at any instant one single current flowing around the complete circuit in the manner indicated by the arrows in the diagram. This current reverses direction at the instant at which all the coils pass the midway points between the *N* and *S* poles. The number of alternations per second is equal to the number of poles multiplied by the number of revolutions per second. The field magnets *N* and *S* of such a dynamo are usually excited by a direct current from some other source.

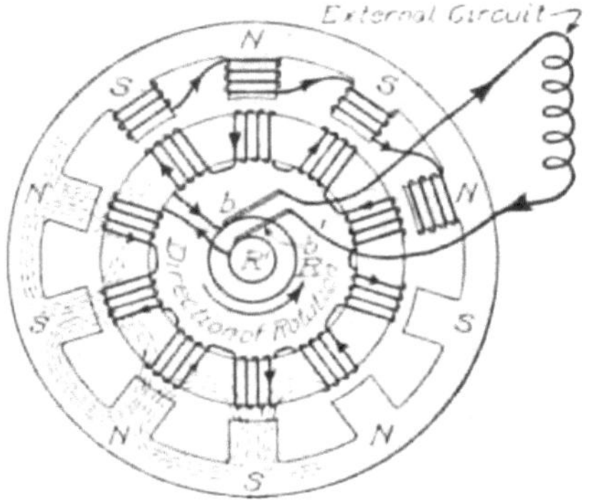

Fig. 216. Diagram of Revolving-Armature Alternator

If, instead of rotating the armature inside the field, we rotate the field inside of the armature, the same result will be obtained. How this is accomplished is shown diagrammatically in Fig. 217 and a commercial revolving-field alternator is shown in Fig. 218. Several poles of the rotating field inside of the armature of a multipolar alternator are shown in Fig. 219.

184. Cycle, Frequency, and Period of Alternating Current. The electromotive force of a loop or a coil of wire rotating between two poles as we have seen produces a set of + and − values during each revolution, Fig. 184. This complete set of + and − values of the e.m.f. produced in a coil during one revolution (360 degrees) of a bipolar machine is called a *cycle*. In the same time, the current alternates twice, so that for any machine the *number of cycles per second is equal* to *one-half the number of alternations per second.* In the case of the multipolar alternators, Figs. 216 to 219, the *cycle* will take place while one coil is moving

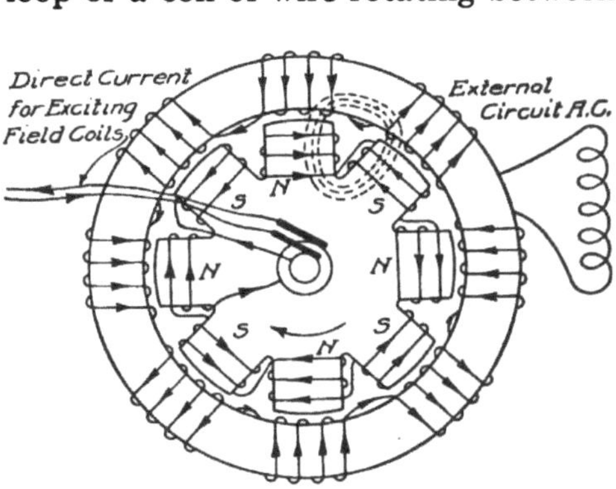

Fig 217. Diagram of Revolving-Field Alternator

from underneath one N to the position underneath the next N pole as in doing so the e.m.f. passes through the same set of values as it does for a complete revolution in a bipolar machine. Hence the rotation from one N pole to the next N pole, Fig. 184, is called 360 *electrical degrees.*

The *frequency* of an alternating current is the number of *cycles per second* through which the current passes and is therefore equal

Fig. 218. Typical Revolving-Field Alternator
Courtesy of Electric Machinery Company

to the number of poles, both N and S, which the coil passes per second, divided by two. The frequency is also stated often as the number of alternations per second, which is equal to the number of poles, both N and S, which the coil passes per second. Thus a frequency of 60 cycles per second, and one of 120 alternations per second are the same. It is, however, better usage to state the frequency in cycles per second. The frequency used on lighting circuits is about 60 cycles per second, while for power circuits they are often as low as 15 cycles per second.

If we let f be the frequency in cycles per second and T be the time in seconds required to complete one cycle, then

$$T = \frac{1}{f}$$

so that the period T of a 60-cycle current is one-sixtieth of a second.

185. Single-Phase Circuit. If the armature shown in Fig. 220 be rotated inside a field having a pole corresponding to each coil on

Fig. 219. Close View of Armature and Field Coils of G. E. Alternator

the armature, the conditions will be the same as in Figs 216 to 219 with regard to the character of the induced e.m.f. The induced currents in all of the coils at any instant will be in the same direction and all will reach their maximum and minimum values at the same time, and are therefore added to each other. The resulting current in the external circuit is a *single-phase* alternating current.

186. Polyphase Circuit. If, instead of having the number of coils on the armature equal to the number of poles on the field, as in the case of the single-phase alternator, we have three coils on the armature to one pole on the field as in Fig. 221, we can obtain three separate currents from the alternator by connecting up the coils of the armature in the following manner:

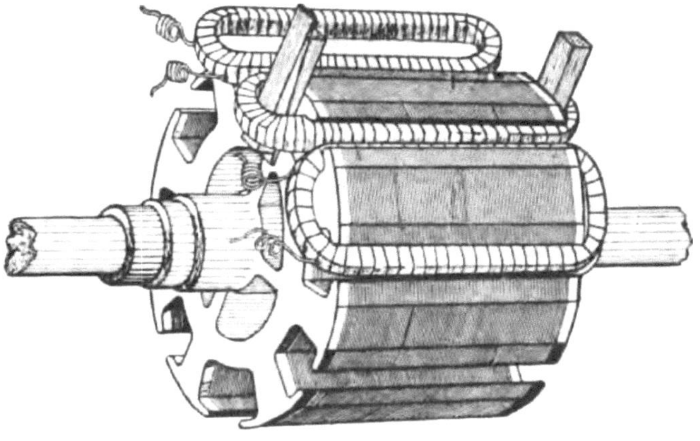

Fig. 220. Coils Mounted on Rotor of Single-Phase Alternator

Let all of the windings marked a_1, a_2, a_3, etc., be connected in series to one pair of collecting rings; the wires in a_1 would pass up through the slot, in a_2 down through the slot, in a_3 up, etc. Likewise in slot b_1 the wires pass up, in b_2 down, etc., and in c_1 up, in c_2 down, etc. In all, six collecting rings are required if the three currents A, B, and C are to be entirely independent of each other. The three currents A, B, and C will not reach their maximum values at the same time but will be 120 electrical degrees apart, as shown by the curves in Fig. 222. The curve marked A is the sum of the induced electromotive forces in the wires which are contained in the slots, a_1, a_2, a_3, etc., of the armature. The B curve represents the resultant e.m.f. induced in the wires in the slots by b_1, b_2, b_3, etc., and similarly for the C curve. Fig. 222 represents the

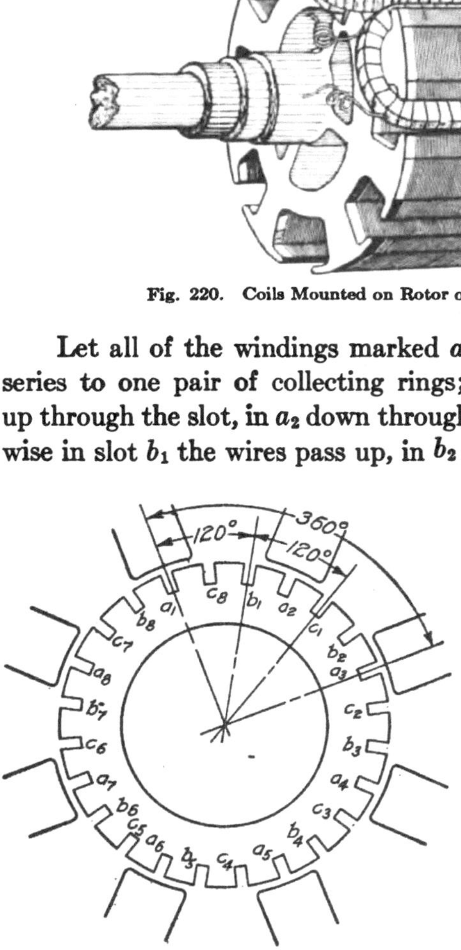

Fig. 221. Arrangement of Coils in Polyphase Alternator

conditions in the three circuits while a_1 is rotating from the position shown in a clockwise direction to the position occupied by a_5, that is, through 720 degrees or two cycles.

If we take a_1 as passing through a positive maximum to start with, as shown in Figs. 221 and 222, and remember that the wires

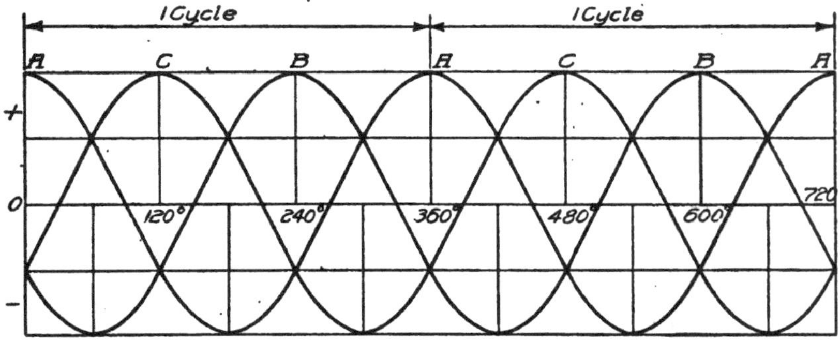

Fig. 222. E.M.F. Curve for Three-Phase Alternator

pass up through the slots with odd subscripts and down through the slots with even subscripts, then when b_8 has reached the position of a_1 in Fig. 221, a rotation of 60 degrees, the induced e.m.f. will be a maximum but will be negative, since the wires in this slot pass down instead of up. Thus the curve *B* reaches a maximum negative value at this instant. A further rotation of 60 degrees brings c_7 to the position occupied by a_1 and it will have a + maximum induced in it and likewise all of the *c* coils so that the *C* curve reaches a + maximum 120 degrees after the *A* curve reaches it. Similarly the *B* curve reaches a + maximum 120 degrees farther on in the rotation. The number of electrical degrees between two + maxima on the curves *A* and *C* is 120 which is called their *phase difference*. Thus the three-phase alternator produces three currents which with the same external conditions will differ in phase from each other by 120 degrees, and whose sum at any instant, as will be seen from Fig. 222 is zero. The current *C* has a *lag* of 120 degrees on current *A*, and a *lead* of 120 degrees on current *B*, since

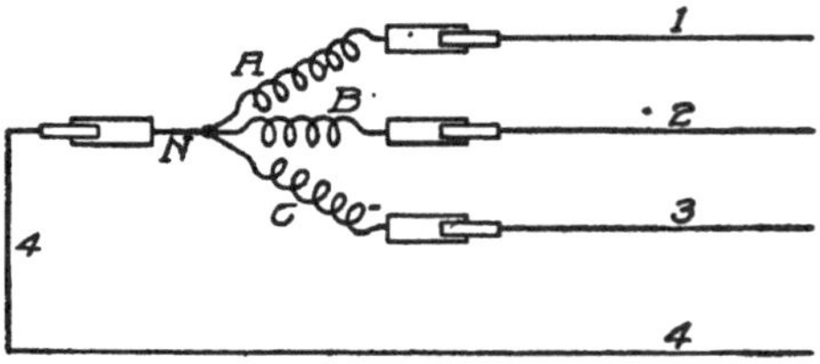

Fig. 223. Collector Ring System for Three-Phase Alternator

it reaches its maximum 120 electrical degrees after *A* does, and 120 electrical degrees before *B* does.

In a three-phase four-wire system the coils are connected as in Fig. 223, the wire numbered 4 serving as a return wire for each of the three mains marked 1, 2, and 3.

It will be remembered that inductance in an a.c. circuit behaved like inertia in mechanics, retarding the current and producing a *lag*. Capacity, on the other hand, produced a *lead* of the current over the e.m.f. In the "balanced system" of wiring for the three-phase alternator the three receiving circuits each have the same resistance, and the same reactance. They will have the same reactance when the inductance and capacity in each of the three circuits is such as to produce currents which are 120 electrical degrees apart, each current

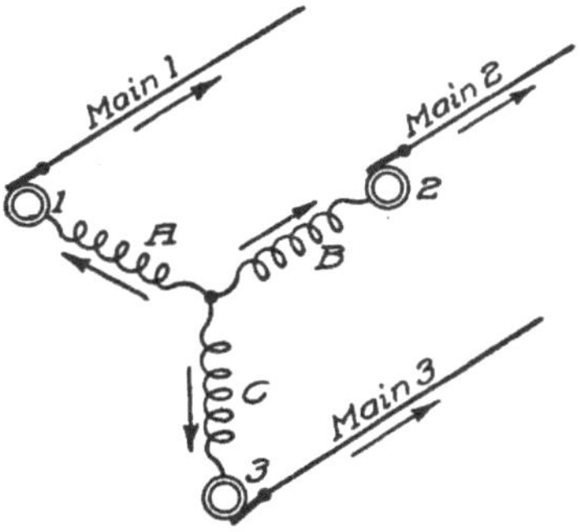

Fig. 224. Y-Diagram for Three-Phase Alternating Circuit

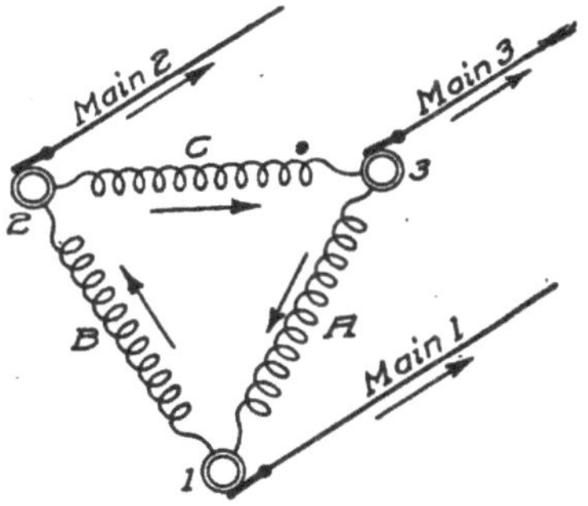

Fig. 225. Δ-Diagram for Three-Phase Alternating Circuit

lagging by the same amount behind the e.m.f. producing it. When a circuit is so balanced, wire 4 may be dispensed with and either the "Y" or "Δ" scheme shown in Figs. 224 and 225 may be used. This system is commonly used for transmitting high-tension currents (currents transmitted at high voltage).

ALTERNATING-CURRENT MOTORS

187. Synchronous Motor. A synchronous motor and an a.c. generator bear the same relation to each other as do the d.c. motor and dynamo. When the a.c. generator is used as a motor, it must first be brought into synchronism with the current supplied to it, that is, it must be in step with the generator. It is therefore not used for ordinary purposes since it is not self-starting. Its main use is for driving d.c. dynamos in substations.

188. Single-Phase-Series Motors. Motors which are supplied with a commutator and may be used on either a.c. or d.c. circuits are now very common. They are used largely for electric-fan motors, vacuum cleaners, and small motors for various purposes. A few street-car lines are equipped with motors which run on either a.c. or d.c. circuits, in which case the frequency of the a.c. used is small.

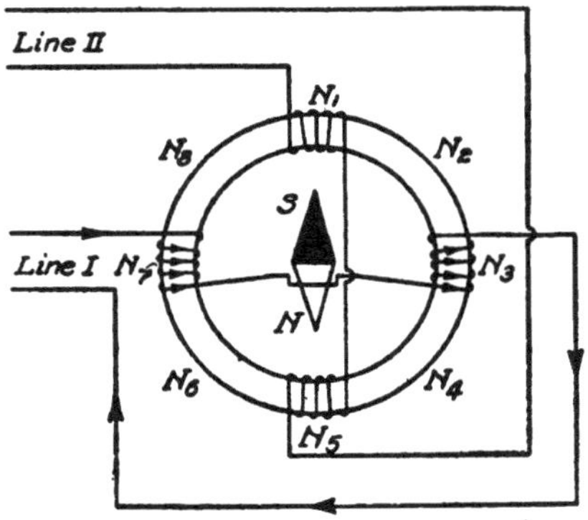

Fig. 226. Principle of Induction Motor. Iron Ring Magnetized by Two Alternating Currents Having Phase Difference of 90 Degrees

189. Induction Motor. The principle of the rotating field of an induction motor can be very strikingly demonstrated by magnetizing an iron ring by passing the current from two a.c. lines around the ring as shown in Fig. 226. The two currents used are of the same frequency but differ in phase by 90 degrees as shown in Fig. 227, the current in line *II* having a *lag* of 90 degrees over that in line *I*, that is, it reaches its maximum 90 degrees farther on in the cycle.

Starting at the beginning of the cycle shown in Fig. 227, the current in line *I* is a maximum and in line *II* is zero. Applying the right-hand helix rule to the coils on line *I*, Fig. 226, we see that the top of

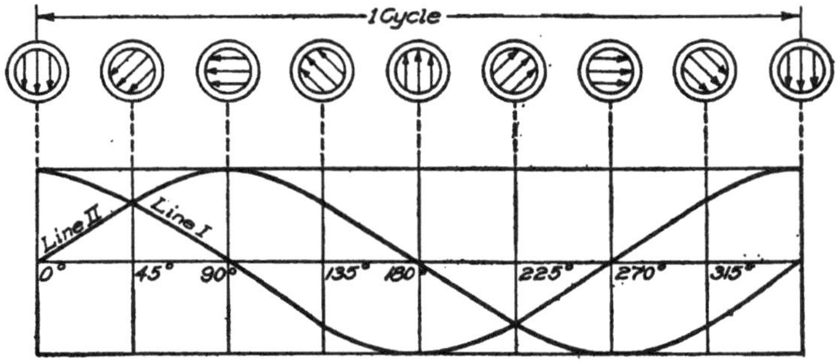

Fig. 227. Rotating Magnetic Field Produced by Two Alternating Currents Having Phase Difference of 90 Degrees

the ring will be made an *N* pole and the bottom an *S* pole, so that the compass inside the ring points to the south. This result is shown diagrammatically by the rings at the top of Fig. 227. At the end of one-eighth of a cycle, the current in line I has decreased and in line II has increased, causing the N pole on the ring to travel from N_1 to N_2, Fig. 226. At the end of one-fourth of a cycle, the current in line I is zero and in line II is a maximum so that the right side of the ring is an N pole and the left side an S pole, and so on. Thus the N pole during one cycle travels around the ring occupying in succession the positions N_1, N_2, ... N_8, and all of the intervening points, i.e., the motion of. N is continuous and not by jumps as was necessary in studying the experiment. Thus the field on the inside

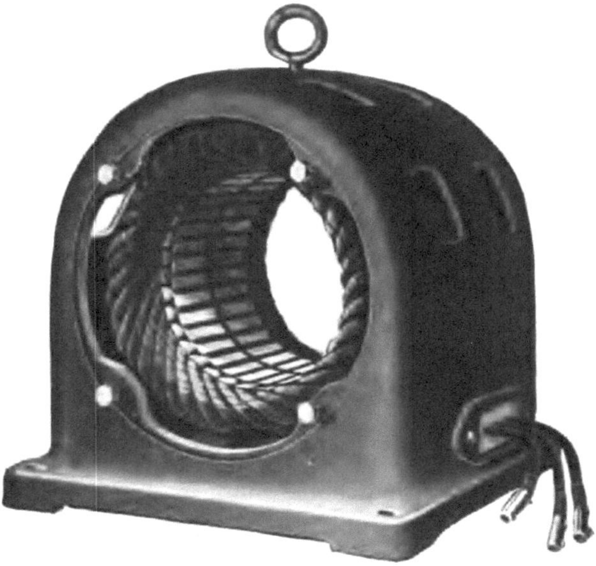

Fig. 228. Stator of Polyphase Induction Motor
Courtesy of Burke Electric Company

is made to rotate continuously through the directions indicated in the circles at the top of Fig. 227, and a magnet on the inside of the ring in Fig. 226 will rotate with the field; in other words, we have a motor of the two-phase type.

A typical commercial induction motor is shown in Figs. 228 and 229. Fig. 228 shows the stationary primary member (stator or field) after the coils have been fastened in place. This produces the rotating field. The secondary member (or rotor) is shown in Fig. 229. The conductors are rectangular bars which are placed in slots around the iron core and project over the ends where they are rigidly attached to massive rings of copper thus forming a short-circuited winding. This type of winding is called a *squirrel-cage rotor*. The rotating field of the stator sweeps across the heavy rectangular-bar windings of the rotor and induces in them strong currents which magnetize the iron core of the rotor and the rotating field drags this magnetized core around just as the rotating field in our model induction motor of Fig. 226 dragged the magnetic needle around. The squirrel-cage induction motor runs at very nearly constant speed for all loads, requires practically no attention, is very simple to operate so that any one can run it, and requires no accessories as does the synchronous motor. For these reasons its use is rapidly increasing for nearly all purposes.

Fig. 229. Rotor of Burke Induction Motor

MEASUREMENT AND TRANSMISSION OF POWER IN AN ALTERNATING-CURRENT CIRCUIT

190. Comparison of A.C. and D.C. Electromotive Forces and Currents. Since the e.m.f. or the current curve for alternating currents is a sine wave with an equal number of + and − values, it is evident that the average e.m.f. or the average current throughout one cycle or any number of cycles is zero, but that during the positive (or negative) part of the cycle, it is not zero. This latter average value is spoken of as the "average value" or "mean value".

If we let i be the value of the alternating current at any instant the heating effect produced will be Ri^2 and the average rate at which heat will be generated during one or more cycles will be R multiplied by the average i^2 (remember this is the average of the squares and not the square of the average). The *effective* value of the alternating current is then equal to the direct current which will produce the same heating effect, or it is equal to $\sqrt{\text{average } i^2}$. This is about .7 of the maximum value of the alternating current. Similarly the effective e.m.f. in an a.c. circuit is about .7 of the maximum e.m.f. Voltmeters and ammeters with alternating currents record the effective values of the e.m.f. and current, respectively.

191. Comparison of Power in A.C. and D.C. Circuits. For the d.c. circuit the power delivered by the circuit in watts is

watts = volts × amperes

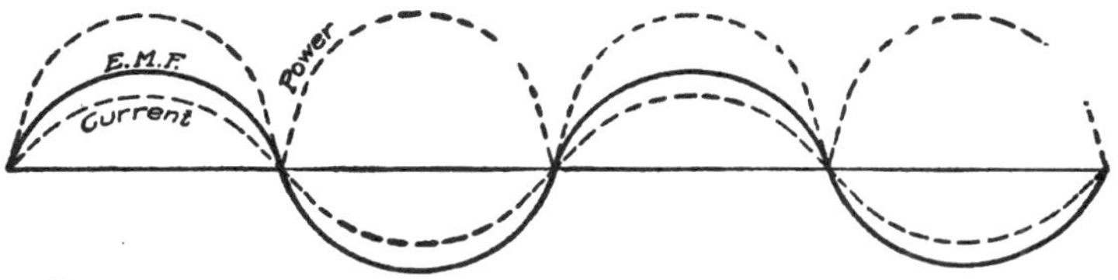

Fig. 230. E.M.F., Current, and Power Curve, for Circuits where E M.F. and Current Are in Phase

For the a.c. circuit, if we let e be the instantaneous value of the e.m.f. and i the corresponding value of the current, then the power delivered by the circuit at that instant in watts is

$$\text{watts} = ei$$

and the average of ei taken over one complete cycle is the "average power," or simply the "power" delivered by the circuit in watts.

For a non-inductive circuit without capacity, the e.m.f. and current will be in phase, in which case the power is given by

watts = effective e.m.f. × effective amperes

The conditions in such a circuit, a lamp for example, are shown in Fig. 230, in which case the watts consumed by the lamp are equal to the a.c. voltmeter reading multiplied by the a.c. ammeter reading. It will be noted that the power curve is always positive.

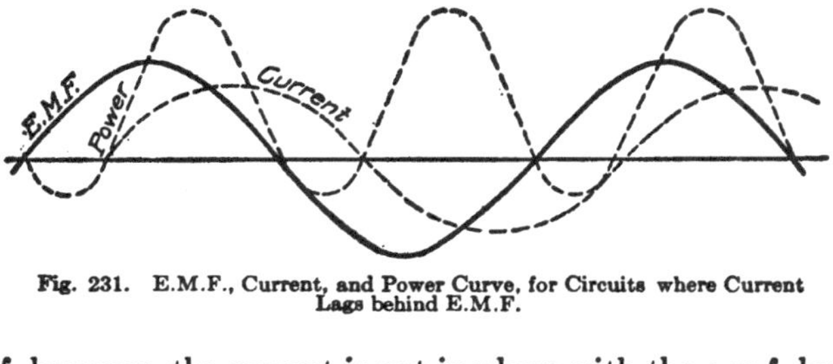

Fig. 231. E.M.F., Current, and Power Curve, for Circuits where Current Lags behind E.M.F.

If, however, the current is not in phase with the e.m.f. but *lags* behind it as in Fig. 231, then the instantaneous values of ei are as indicated by the fine dotted line showing the power curve. It will be seen that this curve is not always positive but is negative for a small part of the time. During this time, the circuit is delivering power to the alternator, and during the time that the power curve is positive, the alternator is delivering power to the circuit. If we let E be the effective e.m.f., I be the effective current, and θ be the number of electrical degrees which the current lags behind the e.m.f., it can be shown that the power P in watts is given by

$$P = E \times I \times \cos \theta$$

Thus, the power consumed by an induction motor (on account of the inductance of the motor, this causes the current to *lag*) when an a.c. voltmeter connected across the terminals reads 110 volts and an a.c. ammeter connected in series in the line reads 5 amperes will *not* be 110×5, or 550 watts, since the current lags. Suppose the current lags 30 degrees behind the e.m.f. in the above example. The

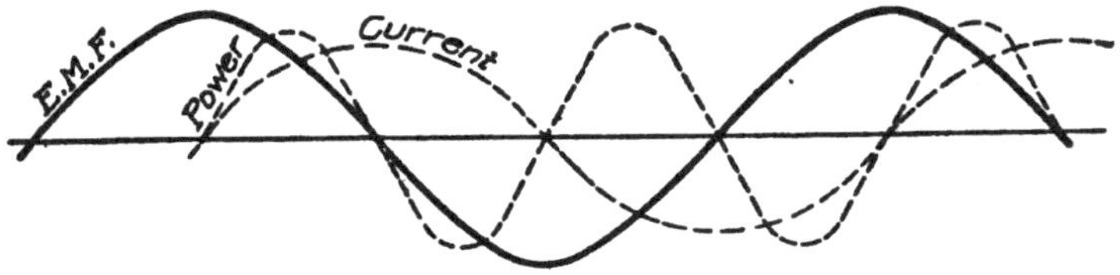

Fig. 232. E.M.F., Current, and Power Curve for Circuits with Large Inductance

cosine of 30 degrees is .866 and hence the power consumed by this motor is

$$P = 110 \times 5 \times .866, \quad \text{or} \quad 476.3 \text{ watts}$$

Suppose we had a circuit in which the current was lagging 90 degrees behind the e.m.f., then $\theta = 90$, and $\cos \theta = 0$, and hence $P = 0$, that is, the circuit would be receiving no power, and therefore the current in the circuit is called a "wattless" current. The inductance of a transformer (Sections 193–199) is so high that when the secondary is open the current lags nearly 90 degrees behind the e.m.f. and hence no power is consumed until the secondary is closed through some circuit. The e.m.f., current, and power curves, for a circuit with very large inductance are shown in Fig. 232. Here the current lags nearly 90 degrees behind the e.m.f. The power curve is plus a trifle more of the time than it is negative, so that the watts consumed by the circuit are very nearly zero.

192. Watt-Hour Meters. It should be kept in mind by the student and user of electricity that wattmeters measure the power in an electrical circuit and that watt-hour meters measure the amount of electrical energy passing through the circuit. Thus if a wattmeter indicated that a lamp circuit was receiving 1 kilowatt, a watt-hour meter left in the same circuit for 30 minutes would indicate .5 of a kilowatt-hour. The kilowatt registered by the wattmeter shows that work is being done on the lighting circuit at the rate of $\frac{4}{3}$ h.p., while .5 of a kilowatt-hour shows that during the 30 minutes $33{,}000 \times 30 \times \frac{4}{3}$, or 1,320,000 foot-pounds of work have been transformed into the energy of the electric current which was registered by the watt-hour meter as .5 of a kilowatt-hour.

The instrument placed in the basement of buildings for measuring the amount of electricity consumed is the watt-hour meter. A

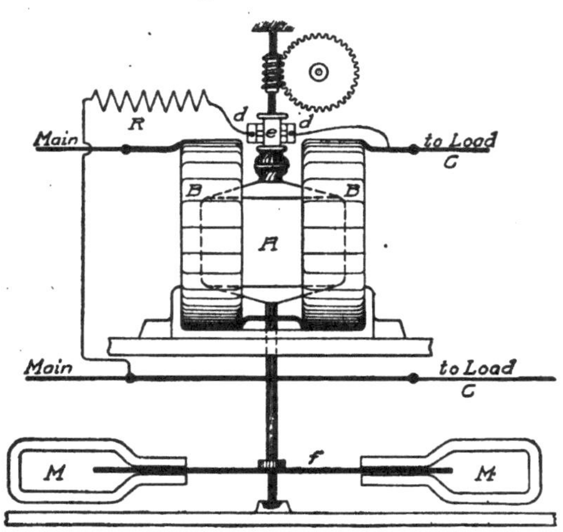

Fig. 233. Diagram Showing Principle of Watt-Hour Meter

type of watt-hour meter, which may be used on either d.c. or a.c. circuits, is shown in Fig. 233. It is built on the principle of the shunt motor and revolves at a rate depending upon the amount of electrical energy passing through it. On the armature shaft is a worm-gear which turns the dials from which the monthly readings are taken for the purpose of computing the electric-light bills.

The field coils *BB* of this motor, Fig. 233, are connected in series with the circuit *CC*, in which the energy to be measured is expended. The armature *A*, together with an auxiliary non-inductive resistance *R*, is shunted across the main circuit *CC*, as shown. Current is led into the armature by means of the brushes *dd* pressing on a small silver commutator *e*. Hence the moving force (torque) acting on the armature is proportional to the product of volts multiplied by amperes, that is, it is proportional to watts. In order that the speed of the armature and hence the reading of the instrument may be proportional to this torque, an aluminum disk *f* is mounted on the armature shaft so as to rotate between the poles of the permanent magnets *M*. The induced eddy currents in this disk set up a field opposing its motion, so that the driving torque is always proportional to the speed. While this instrument may be used for either a.c. or d.c. circuits, it is generally used on d.c. work. For alternating currents a watt-hour meter,* which is in reality a small induction motor, is generally used.

TRANSMISSION OF ELECTRIC POWER

193. Transformer. The commercial transformer is a modified form of the induction coil. The chief difference is that the core *R*, Fig. 234, instead of being straight, is bent into the form of a ring, or is given some other shape such that the magnetic lines of force have a continuous iron path, instead of being obliged to push out into the air, as in the induction coil. Furthermore, it is always an alternating instead of an intermittent current which is sent through the primary *A*. Sending such a current through *A* is equivalent to first magnetizing the core in one direction, then demagnetizing it, then magnetizing it in the opposite direction, etc. The results of these changes in the magnetism of the core is, of course, an induced alternating current in the secondary *B*.

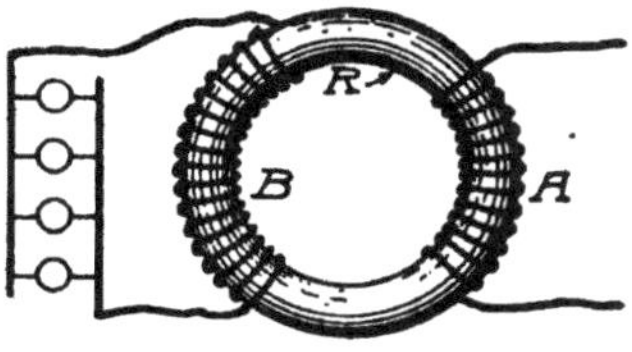

Fig. 234. Diagram of Simple Transformer

194. Use of the Transformer. The use of the transformer is to convert an alternating current from one voltage to another which for some reason is found to be more convenient. For example, in electric lighting where an alternating current is used, the e.m.f. generated by the dynamo is usually either 1,100 or 2,200 volts, a voltage too high to be introduced safely into private houses. Hence, transformers are connected across the main conductors in the manner shown in Fig. 235. The current which passes into the houses to

*See "Induction Watt-Hour Meters" in any standard reference work.

supply the lamps does not come directly from the dynamo. It is an induced current generated in the transformer.

195. Pressure in Primary and Secondary. If there are a few turns in the primary and a large number in the secondary, the transformer is called a *step-up* transformer, because the p.d. produced at the terminals of the secondary is greater than that applied at the terminals of the primary. Thus, an induction coil is a step-up transformer. In electric lighting, however, transformers are mostly of the *step-down* type; i.e., a high p.d., say 2,200 volts, is applied at the terminal of the primary, and a lower p.d., say 110 volts, is obtained at the terminals of the secondary. In such a transformer the primary will have twenty times as many turns as the secondary. In general, *the ratio between the voltages at the terminals of the primary and the secondary is the ratio of the number of turns of wire upon the two.*

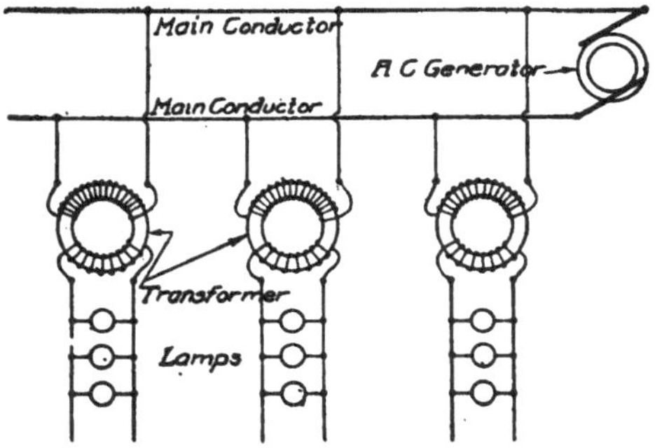

Fig. 235. Alternating Current Circuit Containing Transformers

196. Efficiency of the Transformer. In a perfect transformer the efficiency would be unity. This means that the electrical energy

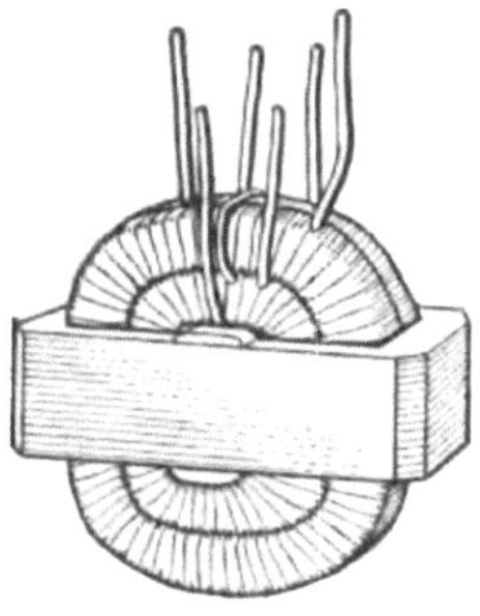
Fig. 236. Transformer Showing Windings and Core

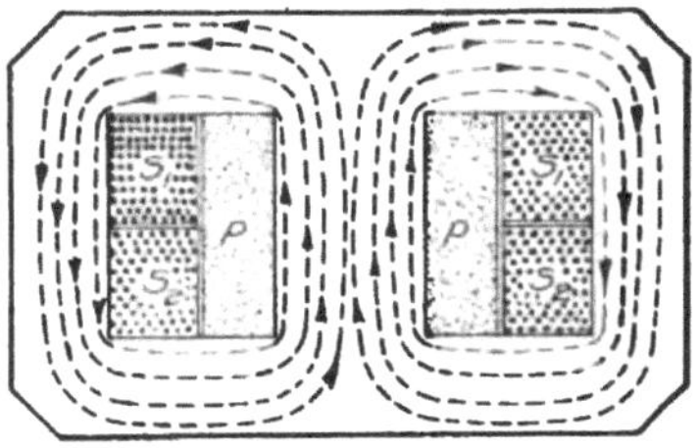

Fig. 237 Cross-Section of Transformer Showing Shape of Magnetic Field

put into the primary, i.e., the volts applied to its terminals, multiplied by the amperes flowing through it, would be exactly equal to the energy taken out in the secondary, i.e., the volts generated in it multiplied by the strength of the induced current; and, in fact, in actual transformers the latter product is often more than 97 per cent of the former, i.e., there is less than 3 per cent loss of energy in the transformation. This lost energy appears as heat in the transformer. This transfer, which goes on in a big transformer, of huge quantities of power from one circuit to another entirely independent circuit, without noise or motion of any sort and almost without loss, is one of the most wonderful phenomena of modern industrial life.

197. Commercial Transformers. Fig. 236 illustrates a common type of transformer used in electric lighting. The core is built

Fig. 238. Typical Transformer Casing

up of sheet-iron laminas about .5 millimeter thick. Fig. 237 shows a section of the same transformer. The closed magnetic circuit of the core is indicated by the arrows. The primary and the two secondaries, which can furnish either 52 or 104 volts, are indicated by the letters p, S_1, and S_2. Fig. 238 is the case in which the transformer is placed. Such cases may be seen attached to poles, Fig. 239, outside of houses in any district where alternating currents are used for electric-lighting purposes.

198. Electrical Transmission of Power. Since the electrical energy produced by a dynamo is equal to the product of the e.m.f.

generated by the current furnished, it is evident that in order to transmit from one point to another a given number of watts, say 10,000, it is possible to have either an e.m.f. of 100 volts and a current of 100 amperes, or an e.m.f. of 1,000 volts and a current of 10 amperes. In the two cases, however, the loss of energy in the wire which carries the current from the place where it is generated to the place where it is used will be widely different. If R represents the resistance of this transmitting wire, the so-called "line", and C the current flowing through it, the heat developed in it will be proportional to C^2R. Hence, the energy wasted in heating the line will be but .01 as much in the case of the high-voltage, 10-ampere current, as in the case of the lower-voltage, 100-ampere current. Hence, for long-distance transmission, where line losses are considerable, it is important to use the highest possible voltages.

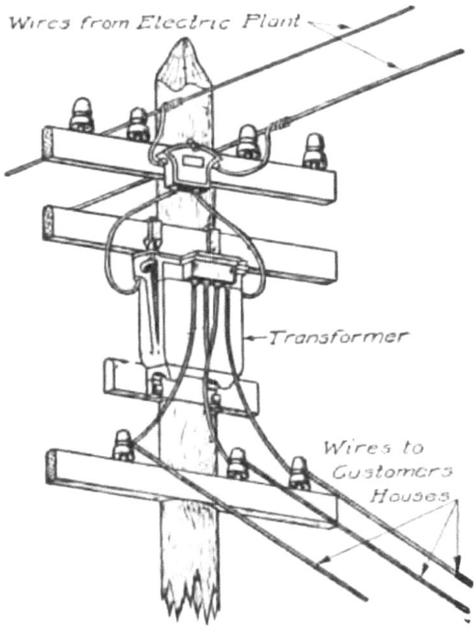

Fig. 239. Transformer Mounted on Electric Light Pole

On account of the difficulty of insulating the commutator segments from one another, voltages higher than 700 or 800 cannot be obtained with direct-current dynamos of the kind which have been described. With alternators, however, the difficulties of insulation are very much less on account of the absence of a commutator. The large 10,000-horsepower alternating-current dynamos on the Canadian side of Niagara Falls generate directly 12,000 volts. This is the highest voltage thus far produced by generators. In all cases where these high pressures are employed, they are transformed down at the receiving end of the line to a safe and convenient voltage (from 50 to 500 volts) by means of step-down transformers.

199. Long-Distance Transmission of Power. It will be seen from the above facts that only alternating currents are suitable for long-distance transmission. Plants are now in operation which transmit power over 100 miles and use pressures as high as 100,000 volts. In all such cases *step-up* transformers, situated at the power house, transfer the electrical energy developed by the generator to the line, and *step-down* transformers, situated at the receiving end,

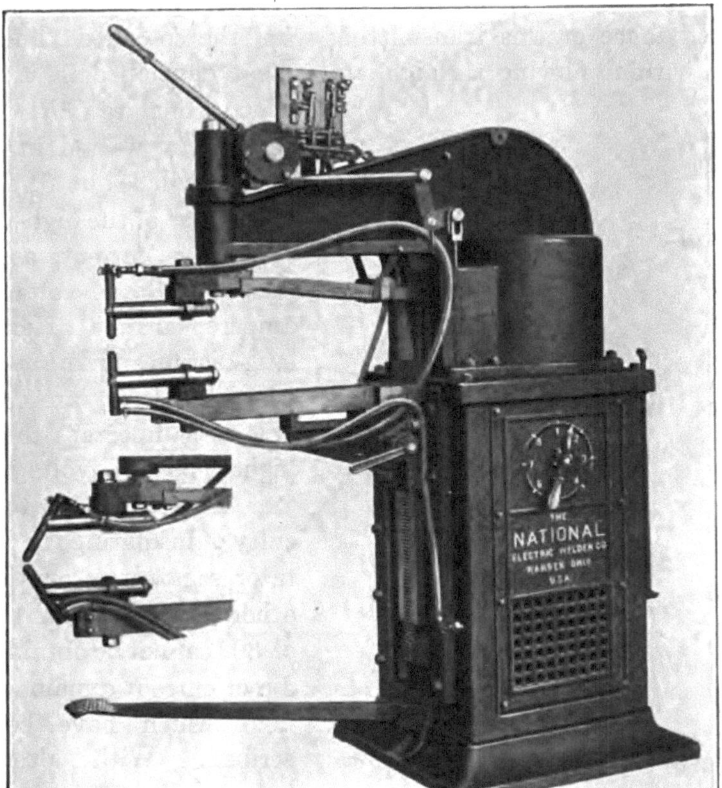

Fig. 240. Typical Electric Spot Welder
Courtesy of National Electric Welder Company

transfer it to the motors, or lamps, which are to be supplied. The generators used on the American side of Niagara Falls produce a pressure of 2,300 volts. For transmission to Buffalo, twenty miles away, this is transformed up to 22,000 volts. At Buffalo, it is transformed down to the voltages suitable for operating the street cars, lights, and factories of the city. On the Canadian side, the generators produce currents at 12,000 volts, as stated, and this is "stepped-up" for transmission to 22,000, 40,000, and 60,000 volts.

200. Spot Welder. A very extensive application of the step-down transformer has been made in recent years in connection with electrical-welding machines. These are made to operate on 110, 220, or higher voltages, but more commonly on 220 volts. The primary of the transformer is connected across the main line, and from the secondary is drawn the current for operating the welder. The voltage in the secondary is only about 3 to 5 volts and the current sufficient to make the power somewhere between 1 h.p. for small work and 200 h.p. for large work. Thus the currents used are very large, a fact which is desirable, of course, since the heat produced is equal to I^2R. A typical spot welder is shown in Fig. 240.

EXAMPLES FOR PRACTICE

1. A multipolar alternator has 20 poles and makes 200 r.p.m. How many alternations per sec. will be produced in the circuit?

2. Two successive coils on the armature of a multipolar alternator are cutting lines of force which run in opposite directions. How does it happen that the currents generated flow through the wires in the same direction? (See Figs. 216 and 217.)

3. What is the period of the alternating current mentioned in Example 1? How many cycles per second in the same current?

4. What happens in the armature of an alternator while it rotates through 360 electrical degrees?

5. What relation exists between the e.m.f.'s induced in the three sets of coils of a three-phase generator? (See Fig. 222.)

6. What is meant by a "balanced system" in connection with a three-phase generator? When the generator is used on a balanced system, how may the coils be connected to the mains to save one wire?

7. What is a synchronous motor?

8. Explain the operation of the induction motor.

9. The watts in an a.c. circuit are not, in general, equal to the product of volts multiplied by amperes. Why not?

10. What relation must exist between the number of turns on the primary of a transformer which feeds 110-volt lamps from a main line whose conductors have a p.d. of 2,300 volts?

11. Why not use a step-up transformer for electrical welders?

INDEX.

INDEX

PAGE

A

B

C

PAGE

PAGE

M

O

P

Made in United States
Troutdale, OR
11/03/2026

56845252R00146